SATELLITE ASTRONOMY:
The Principles and Practice of Astronomy from Space

THE ELLIS HORWOOD LIBRARY OF SPACE SCIENCE AND SPACE TECHNOLOGY
SERIES IN ASTRONOMY
Series Editor: JOHN MASON
Consultant Editor: PATRICK MOORE

In establishing this new series, we aim to co-ordinate a team of international authors of the highest reputation, integrity and expertise in all aspects of astronomy. This series will make a valuable contribution to the existing literature encompassing all areas of astronomical research. The titles in this series will be illustrated with both black and white and colour photographs, and include many line drawings and diagrams, with tabular data and extensive bibliographies. Aimed at a wide readership, the books will appeal to the professional astronomer, undergraduate students, the high-flying 'A' level student, and the non-scientist with a keen interest in astronomy.

PLANETARY VOLCANISM*
PETER CATTERMOLE, Department of Geology, Sheffield University, UK
SATELLITE ASTRONOMY: The Principles and Practice of Astronomy from Space
JOHN K. DAVIES, Royal Observatory, Edinburgh, UK
THE DUSTY UNIVERSE*
ANEURIN EVANS, Department of Physics, University of Keele, UK
SPACE-TIME AND THEORETICAL COSMOLOGY*
MICHEL HELLER, Department of Philosophy, University of Cracow, Poland
ASTEROIDS: Their Nature and Utilization
CHARLES T. KOWAL, Space Telescope Institute, Baltimore, Maryland, USA
ELECTRONIC AND COMPUTER-AIDED ASTRONOMY*
IAN S. McLEAN, United Kingdom Infrared Telescope Unit, Hilo, Hawaii, USA
URANUS: The Planet, Rings and Satellites*
ELLIS D. MINER, Jet Propulsion Laboratory, Pasadena, California, USA
THE PLANET NEPTUNE*
PATRICK MOORE
ACTIVE GALACTIC NUCLEI*
IAN ROBSON, Director of Observatories, Lancashire Polytechnic, Preston, UK
ASTRONOMICAL OBSERVATIONS FROM THE ANCIENT ORIENT*
RICHARD F. STEPHENSON, Department of Physics, Durham University, Durham, UK
THE TERRESTRIAL PLANETS FROM SPACECRAFT*
YURI A. SURKHOV, Chief of the Laboratory of Geochemistry of the Planets, USSR Academy of Sciences, Moscow, USSR
THE HIDDEN UNIVERSE*
ROGER J. TAYLER, Astronomy Centre, University of Sussex, UK
THE ORIGIN OF THE SOLAR SYSTEM*
MICHAEL M. WOOLFSON, Department of Physics, University of York, UK and
JOHN R. DORMAND, Department of Mathematics and Statistics, Teeside Polytechnic, Middlesborough, UK
TO THE EDGE OF THE UNIVERSE*
ALAN and HILARY WRIGHT, Australian National Radio Astronomy Observatory, Parkes, New South Wales

* *In preparation*

SATELLITE ASTRONOMY:
The Principles and Practice of Astronomy from Space

JOHN K. DAVIES, B.Sc., Ph.D.
Astronomer
Royal Observatory, Edinburgh

ELLIS HORWOOD LIMITED
Publishers · Chichester

Halsted Press: a division of
JOHN WILEY & SONS
New York · Chichester · Brisbane · Toronto

First published in 1988 by
ELLIS HORWOOD LIMITED
Market Cross House, Cooper Street,
Chichester, West Sussex, PO19 1EB, England
The publisher's colophon is reproduced from James Gillison's drawing of the ancient Market Cross, Chichester.

Distributors:

Australia and New Zealand:
JACARANDA WILEY LIMITED
GPO Box 859, Brisbane, Queensland 4001, Australia

Canada:
JOHN WILEY & SONS CANADA LIMITED
22 Worcester Road, Rexdale, Ontario, Canada

Europe and Africa:
JOHN WILEY & SONS LIMITED
Baffins Lane, Chichester, West Sussex, England

North and South America and the rest of the world:
Halsted Press: a division of
JOHN WILEY & SONS
605 Third Avenue, New York, NY 10158, USA

South-East Asia
JOHN WILEY & SONS (SEA) PTE LIMITED
37 Jalan Pemimpin # 05–04
Block B, Union Industrial Building, Singapore 2057

Indian Subcontinent
WILEY EASTERN LIMITED
4835/24 Ansari Road
Daryaganj, New Delhi 110002, India

© 1988 J. K. Davies/Ellis Horwood Limited

British Library Cataloguing in Publication Data
Davies, John Keith, *1955*–
Satellite astronomy.
Astronomical bodies. Observation from space vehicles
I. Title
522'.6

Library of Congress Card No. 88-21955

ISBN 0–7458–0232–X (Ellis Horwood Limited)
ISBN 0–470–21174–1 (Halsted Press)

Printed in Great Britain by Hartnolls, Bodmin

COPYRIGHT NOTICE
All Rights Reserved. No part of this publication may be reproduced, stored in a retrieval system, or transmitted, in any form or by any means, electronic, mechanical, photocopying, recording or otherwise, without the permission of Ellis Horwood Limited, Market Cross House, Cooper Street, Chichester, West Sussex, England.

For Maggie and the children

Table of contents

Preface xi

Acknowledgements xii

1 **Astronomy from space, why and how?** **1**
   1.1 Introduction 1
   1.2 Astronomy from balloons 4
   1.3 Astronomy from aircraft 6
   1.4 Astronomy from rockets 7
   1.5 Astronomy from satellites 10
   1.6 Developing an astronomical satellite 12
   1.7 The choice of orbit 17
       1.7.1 Communications 17
       1.7.2 Background radiation 19
       1.7.3 Sky coverage 21
       1.7.4 The effects of residual atmosphere 23
   1.8 Some other design considerations 24
       1.8.1 Attitude control systems 24
       1.8.2 Cooling of scientific instruments 25
       1.8.3 Design flexibility 27
   1.9 Scientific operations 28
   1.10 Conclusion 29

2 **X-ray astronomy** **30**
   2.1 Introduction 30
   2.2 X-rays from the sun 31
   2.3 The first X-ray 'stars' 33

Contents

- 2.4 X-ray surveys 34
  - 2.4.1 Initial attempts 34
  - 2.4.2 SAS-1 (Uhuru) 38
  - 2.4.3 Other early X-ray experiments 41
  - 2.4.4 Ariel 5 45
  - 2.4.5 Other small X-ray missions 48
  - 2.4.6 HEAO-1, end of an era 50
- 2.5 The X-ray observatories 53
  - 2.5.1 HEAO-2 (Einstein) 53
  - 2.5.2 Small observatories 69
  - 2.5.3 EXOSAT 61
  - 2.5.4 X-ray imaging at high energies 65
  - 2.5.5 The Kvant module 67
  - 2.5.6 Astro-C 'Ginga' 69
- 2.6 Future X-ray astronomy missions 69

3 Gamma ray astronomy 72
- 3.1 Introduction 72
- 3.2 Detection methods 73
  - 3.2.1 Scintillation counters 73
  - 3.2.2 Solid state detectors 74
  - 3.2.3 Spark chamber telescopes 75
- 3.3 Early experiments 76
- 3.4 Gamma ray bursters 78
  - 3.4.1 The Vela programme 78
  - 3.4.2 Other gamma ray burst detectors 81
  - 3.4.3 The source of the gamma ray bursts 82
- 3.5 The surveys 83
  - 3.5.1 SAS-2 83
  - 3.5.2 COS-B 85
  - 3.5.3 The gamma ray sky 86
- 3.6 HEAO-3 and gamma ray spectroscopy 87
- 3.7 Future gamma ray missions 88
  - 3.7.1 GAMMA-1 88
  - 3.7.2 GRANAT 89
  - 3.7.3 The Gamma Ray Observatory (GRO) 91

4 Ultraviolet and extreme ultraviolet astronomy 94
- 4.1 Introduction 94
- 4.2 Rocket observations 95
- 4.3 Early ultraviolet observations from satellites 96
- 4.4 The orbiting astronomical observatories 96
  - 4.4.1 OAO-1 97
  - 4.4.2 OAO-2 97
  - 4.4.3 OAO-B 99
  - 4.4.4 OAO-3 Copernicus 100
- 4.5 TD-1A 103

Contents

4.6	Other ultraviolet experiments	105
	4.6.1 Apollo 16	105
	4.6.2 Skylab	106
	4.6.3 Orion 1 and 2	108
	4.6.4 Astronomische Nederlandse Satelliet (ANS)	109
	4.6.5 D2B-AURA	110
	4.6.6 The Galactika experiment on Prognoz-6	110
4.7	The International Ultraviolet Explorer (IUE)	110
4.8	Astron-1	117
4.9	Spacelab-1	118
4.10	The Glazar Telescope on Mir	118
4.11	Future missions	118
	4.11.1 The Hubble Space Telescope	118
	4.11.2 ASTRO-1	119
	4.11.3 LYMAN	120
4.12	Far ultraviolet astronomy	120
4.13	Extreme ultraviolet (EUV/XUV) astronomy	121
	4.13.1 The ASTP EUV telescope	121
	4.13.2 Future missions	123

5 Optical astronomy from space 127
5.1	The Hubble Space Telescope	127
	5.1.1 Introduction	127
	5.1.2 The HST spacecraft	130
	5.1.3 The scientific instruments	131
	5.1.4 Operating the HST	133
	5.1.5 Servicing the HST	133
5.2	Hipparcos	137

6 Infrared and millimetre astronomy from space 143
6.1	Introduction	143
6.2	The surveys	145
	6.2.1 The 2.2 micron and AFGL surveys	145
	6.2.2 The Infrared Astronomical Satellite (IRAS)	146
	6.2.3 The Spacelab-2 Infrared Telescope (IRT)	156
6.3	The observatories	158
	6.3.1 The German Infrared Laboratory (GIRL)	158
	6.3.2 The Infrared Space Observatory (ISO)	158
	6.3.3 The Space Infrared Telescope Facility (SIRTF)	161
6.4	Studying the cosmic background radiation	161
	6.4.1 Introduction	161
	6.4.2 RELIKT-1 and 2	162
	6.4.3 The Cosmic Background Explorer (COBE)	163
6.5	Millimetre and submillimetre astronomy	166
	6.5.1 Salyut 6 BST-1M	166
	6.5.2 Aelita	166
	6.5.3 The Far Infrared Space Telescope (FIRST)	167
	6.5.4 The Large Deployable Reflector (LDR)	167

Contents

7 Radio astronomy from space **168**
   7.1   Introduction 168
   7.2   Second generation radio astronomy experiments 169
          7.2.1   Radio Astronomy Explorer-1 169
          7.2.2   Explorer 43 172
          7.2.3   Radio Astronomy Explorer-2 172
          7.2.4   Salyut 6 KRT-10 (Cosmic Radio Telescope-10) 175
   7.3   Space VLBI 175
          7.3.1   Experiments with the TDRSS satellite 176
          7.3.2   Future Space VLBI missions 177

8 The future **180**
   8.1   Introduction 180
   8.2   The space station 181
   8.3   In-orbit assembly 183
   8.4   Satellite clusters 186
   8.5   Astronomy from the Moon 187
   8.6   Conclusion 188
        Bibliography 189
        Index 194

Preface

It is common to find, especially in popular publications, remarks such as 'IRAS discovered this' or that the 'International Ultraviolet Explorer confirmed that', but these statements are only approximations to the truth. Satellites do not discover anything; people make discoveries. Like telescopes, microscopes, and hundreds of other devices, satellites are tools which make discoveries possible.

However, lest I be thought pedantic, there is some merit in such statements, for they reflect the fact that no one person is responsible for any discovery made with a satellite; behind each astronomer stands a small army of engineers, designers, programme managers, and administrators. There is even an army of ghosts, men and women who have joined a project, made a contribution, and left again long before 'their' satellite reached space. Regarded in this light it is fair to say that 'satellite X discovered', for by saying so we acknowledge the people without whom the satellite would not exist. It is in this sense that I have referred to discoveries 'made' by satellites in this book; the credit belongs not to the hardware, but to the people who made it possible. I should like to dedicate this book to all the men and women who have contributed to the achievements described in these pages.

John Davies
April 1988

Acknowledgements

I would like to thank the following individuals and organisations who have helped me by supplying information and photographs, or by reading and commenting on sections of the manuscript. I should also acknowledge the many astronomers and engineers who, in the normal course of their work, have supplied me with essential background information, and to my colleagues on the IRAS, ROSAT, and ISO projects who have helped me to share in the excitement of doing astronomy from space. Without them this book could not have been written.

Aerospatiale, Dr J. Albinson, Ball Aerospace Systems, Bell Telephone Laboratories, Dr K Bennett, Boeing Aerospace Company, British Aerospace, Dr J. Burnell, Center for Astrophysics—Smithsonian Astrophysical Observatory, CNES (French Space Agency), Dr M. Coe, Dr A. J. Dean, Dornier Systems, ESA ESTEC, ESA Paris, Prof. G. Fazio, General Electric Company, Dr P. K. S. Harper, Dr A. Harris, Dr H. Hirabayashi, ISAS (Japanese Scientific Space Agency), Dr D. Koch, Lockheed Missiles and Space Company, Los Alamos National Laboratory, Dr P. Mandrou, Marconi Space Systems, Martin–Marietta Corporation, MATRA, MBB, NASA Goddard Spaceflight Center, NASA HQ Washington, NASA Marshall Spaceflight Center, NASA Ames Research Center, Naval Research Laboratory Washington, Perkin-Elmer Corporation, Dr J. Pye, Space Research Dept—University of Birmingham, Space Science Laboratory—University of California Berkeley, Space Telescope Science Institute, Dr J. Terrell, TRW Inc, Prof. M. Ulmer.

Part of Chapter 5 is based on material which originally appeared in *Spaceflight*, the magazine of the British Interplanetary Society.

1
Astronomy from space, why and how?

1.1 INTRODUCTION

The atmosphere has always been an uncertain barrier to astronomers. Casual star gazers are often frustrated by clouds, and for professional astronomers the atmosphere is a serious handicap; even on a clear night the air above our heads blocks most of the radiation which arrives from space, and distorts what little information does reach the ground. To overcome these difficulties astronomers have moved their instruments steadily higher, first to observatories located on mountains, then to cages suspended below balloons, and then, albeit briefly, above the atmosphere on rockets. The discoveries which flowed from these short rocket flights led to the development of automatic observatories in orbit around the Earth, and these instruments opened new windows on the sky and revolutionised the way in which mankind views the Universe. Like the primitive telescopes of Galileo, satellites have propelled astronomy into a new age of discovery.

The need to lift astronomical instruments high above the ground stems from the fact that although most of the information that reaches the Earth from space comes in the form of electromagnetic waves, the atmosphere is opaque to almost all this radiation. For centuries astronomers were restricted to observing in the narrow range of wavelengths (about 400 to 700 nm) to which the human eye responds, and even the development of photographic emulsions sensitive to a wider range of wavelengths could not extend the view of the sky beyond the 290 to 1000 nm limits set by the transmission of the atmosphere. The discovery, in 1932, of radio waves from the centre of our Galaxy led to the opening of a new window of wavelengths from about 1 millimetre to several metres, but, despite the achievements of optical and radio astronomers, the fundamental barrier of atmospheric absorption at other wavelengths remained.

Table 1.1 shows the labels which astronomers attach to radiation of different wavelengths. Note that at the gamma and X-ray end of the scale the emphasis is placed

Table 1.1
The electromagnetic spectrum

Wavelength (metres)	Other common units		Photon energy	Frequency	Usual name	Produced by temperatures of approximately (degrees absolute)	Typical objects of astronomical interest
Shorter than			*Greater than*				
10^{-13}			80.6 MeV		⎱ Gamma ray		Cosmic ray interactions with interstellar gas
10^{-12}			8.06 MeV		⎰		
10^{-11}			0.8 MeV				
10^{-10}	1 Å,	0.1 nm	80.6 KeV		Hard X-ray	100 000 000	Hot gas in clusters of galaxies
10^{-9}	10 Å,	1 nm	8.06 KeV		Soft X-ray	10 000 000	Accretion disks in binary systems
10^{-8}	100 Å,	10 nm	0.806 KeV		XUV/EUV/Far UV	1 000 000	White dwarf stars / Flare stars
10^{-7}	1000 Å,	100 nm	0.08 keV		Ultraviolet	100 000	O type stars
3×10^{-7}		200 nm			Violet ⎱		
4×10^{-7}		400 nm	⎱ Optical window		⎰ Range of human eye		
7×10^{-7}		700 nm	⎰		Red ⎰		
8×10^{-7}	0.8 μm‡				Near infrared	10 000	K, M stars
10^{-6}	1 μm						
10^{-5}	10 μm				Infrared	1000	Circumstellar dust shells. Comets and asteroids. Cool interstellar dust
10^{-4}	100 μm				Far infrared	100	
10^{-3}	1 mm			300 000 MHz	Millimetre	10	Microwave Background
10^{-2}	1 cm			30 000 MHz	Microwave	1	
10^{-1}	10 cm			3 000 MHz			
1	1 metre			300 MHz			
10	10 metre			30 MHz			Electrons spiralling in magnetic fields
10^{2}	100 metre			3 MHz	Short Wave		
10^{3}	1 km			300 kHz	Medium wave		
10^{4}	10 km			30 kHz	Long wave		
10^{5} and greater	100 km			3 kHz	Very long wave/ very low frequency		

Introduction

Table 1.2
Absorption by the atmosphere

Approximate wavelength (metres)	Usual name	Penetrates to (typically)	Absorption mechanism
Energies more than 100 000 MeV	Ultrahard Gamma ray	40 km	Atomic processes. Detectable at ground level indirectly via showers of sub-atomic particles.
10^{-13} to 10^{-12}	Gamma rays with energies greater than a few MeV	40 km	Pair production. Gamma ray decays to electron and positron as it passes close to an atomic nucleus in the atmosphere.
10^{-11} to 10^{-10}	Low energy gamma rays and very hard X-rays	40 km	Compton scattering. Photon knocks out an electron from an atom of atmospheric gas, loses energy, and is shifted to a longer wavelength.
10^{-9}	Hard X-ray	70–100 km	Photoelectric effect. Photon knocks electron out of atom of atmospheric gas. Energy of photon is transferred to electron.
10^{-8}	Soft X-ray/XUV/EUV	150 km	
10^{-7} to 3×10^{-7}	Ultraviolet	50–100 km	Dissociation and ionisation of atmospheric molecules such as oxygen, nitrogen, and ozone.
4×10^{-7} to 7×10^{-7}	Visible	—	Not absorbed, but scattering and image degradation occurs.
10^{-6}	Near infrared		Absorption by carbon dioxide and water vapour. Photons have enough energy to cause bending and rotation of molecules. Some atmospheric windows from 1–4 μm, around 10 and 20 μm, and about 350 μm. High, dry sites required for infrared and submillimetre telescopes.
10^{-5}	Infrared	5–10 km	
10^{-4}	Far infrared		
10^{-3}, 10^{-2}	Millimetre		
10^{-1}, 1, 10	Microwave	—	Radio window, varies with state of ionosphere which itself depends on the amount of solar activity.
10^2, 10^3	Short wave, Medium wave	90–500 km	Radio waves reflected back into space by the ionosphere.
10^4 and greater	Long wave and very long wave/Very low frequency		

not on the wavelength of the radiation, but on the energy carried by each photon. (A photon is a packet of electromagnetic waves, and its energy is defined by the expression hc/λ where h is Planck's constant, c is the speed of light and λ is the wavelength of the radiation. This energy is usually expressed in multiples of electron volts.) It is, however, important to realise that the boundaries between these regions are in many respects arbitrary and that the difference between a low energy gamma ray and a high energy (or hard) X-ray is only the small difference in wavelength of, and hence in the energy carried by, each photon. Table 1.2 illustrates the height to which various types of radiation penetrate, and summarises the processes responsible for their absorption.

1.2 ASTRONOMY FROM BALLOONS

In view of the limits placed by the atmosphere it is not surprising that astronomers have tried to rise above it, and the use of balloons for astronomical purposes probably began with the flight organised in 1874 by Jules Janseen of the Paris observatory at Meudon. Janseen was trying to prove that certain lines in the spectrum of sunlight were produced within the Earth's atmosphere, and sent two aeronauts to record the solar spectrum with a hand held spectroscope in the (vain) hope that the lines would disappear at high altitude. This flight, and another in which two of a three man crew perished during an ascent to over 10 000 metres, were followed by the remarkable efforts of Count Aymar de la Baume-Pluvinel who developed both an automatic photographic spectrograph and a solar homing device for use in an unmanned balloon flight. After a failure in July 1898, when the balloon drifted out of sight and was lost, a repeat of the experiment in July 1889 successfully obtained a spectrum which showed that some of the features in the solar spectrum became weaker when observed from high altitude. This result enabled the Count to show that these features did not arise in the Sun, but were due to the Earth's atmosphere. A third flight later the same year failed, and the Count then abandoned his scientific balloon flights and henceforth restricted himself to carrying his instruments up high mountains. Other experiments were made during the ascents of Auguste Piccard who, together with his brother Jean and sister-in-law Jeanette, made a number of high altitude (20 km) flights in the 1930s. Some of these flights carried instruments to investigate the properties of cosmic rays, but, despite these early efforts, it was not until the 1950s that astronomical ballooning developed in a major way.

In 1954 Audouin Dollfus, like Janseen also from the Meudon observatory, was carried in a balloon flown to 7000 metres by his father and took spectra of Mars during a search for water vapour in the Martian atmosphere. In 1956 and 1957 high quality photographs of the Sun were taken by Dollfus, Donald Blackwell, and D. W. Dewhirst from a balloon which carried them and a 26.5 cm $f/12$ refracting telescope to an altitude of 6 km. Although Dollfus made another ascent in 1959, manned balloons were soon replaced by higher flying unmanned experiments like the Princeton University Stratoscope I telescope which could lift a 30 cm, $f/8$ telescope to an altitude of around 25 km. Stratoscope I made five flights, three in 1957 (22 August, 25 September, and 17 October) and two in 1959 (11 July and 17 August) during which photographs of the Sun showing sunspots and the solar granulation in previously unseen detail were obtained. The success of Stratoscope I led to a number of similar projects, including the launch of instruments designed to observe the solar corona. One of these, Coronascope I, carried a small $f/10$ coronagraph to an altitude of around

25 km on two flights in 1960 (10 September and 3 October), while a third flight, using improved equipment and named Coronascope II, reached 30 km in March 1964. Other balloon borne instruments reached even higher; on 20 July 1963 a University of Minnesota payload of solar experiments was carried 33.5 km above the Earth.

Stratoscope I was followed by an ambitious attempt by Princeton University and the Perkin Elmer corporation to develop a multipurpose, unmanned balloon observatory known as Stratoscope II. This instrument, carrying an 86 cm, $f/4$ telescope to an altitude of 26 km, was able to lock onto and follow a star with an accuracy of better than one arc second, while the telescope and its scientific instruments were operated by a radio link from the ground. The first flight, on 1 March 1963, carried infrared instruments which were used to take spectra of Mars; and the second mission, on 26 November 1963, observed the near infrared spectra of nine red giant stars. Subsequent flights were used to take high resolution photographs on 70 mm film which was removed from the payload after each flight. The project was terminated after eight

The Stratoscope II balloon borne telescope (Perkin-Elmer)

missions when Stratoscope II struck a tree on landing and suffered irreparable damage to the main mirror of the telescope.

Space precludes a detailed discussion either of Stratoscope II or of many of the other balloon experiments which have followed it. Some of these have carried X-ray and gamma ray detectors high enough to observe energetic photons from space; others have lifted infrared instruments above the absorption produced by atmospheric water vapour and carbon dioxide. Although the pioneering period of balloon astronomy is now over, balloons have not yet reached the end of their usefulness. In particular, the ability to make a flight at fairly short notice, and to apply the lessons learned to a follow up mission a few months later, still makes ballooning a cost effective tool for certain types of astronomical research.

The flexibility of balloon borne instrumentation was demonstrated convincingly by observations of supernova 1987A, an exploding star in the Large Magellanic Cloud. Astronomers had predicted that the peak intensity of gamma ray emission from the supernova would occur in early January 1988, but no satellites equipped with suitable detectors were in orbit at the time. To take advantage of the unique opportunity presented by the supernova, an advanced gamma ray detector, originally planned for a flight on the then grounded Space Shuttle, was transported to Antarctica for a balloon launch. The two tonne payload, mounted in a solar powered gondola and cooled by a supply of liquid nitrogen, was lifted to an altitude of 35 km on January 8. The experiment operated for 72 hours (not all of which was spent gathering data) before a power failure caused the mission to be terminated. Other balloon launches to study the supernova were made from Alice Springs in Australia, including a mission carrying another type of gamma ray detector which was launched in November 1987.

Apart from special operations like those mounted to study supernova 1987A, the world's main balloon launching base today is the National Scientific Balloon Facility (NSBF) in Palestine, Texas. The NSBF flies balloons for research in atmospheric science, astronomy, and cosmic rays, and for engineering development. The balloons float to altitudes of about 40 km, and on occasions make flights lasting many days. Typical payloads may weigh up to two tonnes and are usually recovered by parachute at the end of each mission. The extremely fragile balloons make only one flight, but the payloads are often refurbished and flown again.

1.3 ASTRONOMY FROM AIRCRAFT

Although a high flying aircraft is not as stable a platform as a balloon or satellite observatory, the advent of modern computers and advanced stabilisation systems has made airborne astronomy a viable technique for certain types of research, notably in infrared astronomy. A modern jet transport can carry a telescope, plus a cadre of astronomers and technicians, to altitudes of 12 km and return them safely for re-use time and time again.

Although solar observations were made from an RAF Canberra jet bomber in 1957 and a Concorde carried scientists on a high altitude trip to observe a solar eclipse in the 1970s, the leader in the field is undoubtedly the airborne astronomy branch of NASA. After trials using an A-3B bomber in 1965, NASA began a programme of airborne astronomy with instruments carried in a modified Convair 990 passenger jet. This was later joined by a Lear Jet executive aircraft adapted to carry a 30 cm infrared telescope which looked out of a hole cut in the side of the aircraft's fuselage. The Lear Jet was able to operate at an altitude of 15 km, well above most of the atmospheric water vapour, and so could observe at infrared wavelengths inaccessible from the ground.

The largest airborne telescope in the NASA fleet is presently the 91 cm instrument carried in the Gerard P Kuiper Airborne Observatory (KAO). The KAO is a considerably modified Lockheed C-141 Starlifter military transport which has a hole in the fuselage roof just in front of the wing through which the telescope is pointed. Porous spoilers placed ahead of the opening minimise turbulence and limit pressure changes in the open cavity in which the telescope is located. The telescope is mounted on four shock absorbers, and the entire assembly is balanced on a spherical air bearing which effectively eliminates any vibrations induced by the aircraft. The telescope is stabilised by a combination of the aircraft's own control system plus a set of gyroscopes and star trackers linked to the motors which drive and guide the telescope. Even in conditions of moderate turbulence the result is a telescope which can be guided with an accuracy of about 2 arc seconds. Light from the telescope is sent, via a mirror, to a mounting flange where the scientific instruments are attached. The flange is cut into the wall of a pressurised cabin so that the instrument, and its accompanying team of astronomers, can be carried in relative comfort in an environment similar to that of a mountain top observatory.

Space precludes a detailed discussion of either the KAO or the proposed Stratospheric Observatory For Infrared Astronomy (SOFIA), a 3 metre telescope mounted in a modified Boeing 747 Jumbo Jet. References to further reading on the KAO are given in the *Bibliography*.

1.4 ASTRONOMY FROM ROCKETS

Despite the successes achieved by airborne telescopes, the altitudes at which they operate is not sufficient to open up the electromagnetic spectrum completely. To carry an instrument high enough to observe at some wavelengths, astronomers need a vehicle able to rise well above the 40 km ceiling of a helium filled balloon. The obvious, indeed the only, such vehicle is the rocket.

At the end of the Second World War the Allied forces captured a number of V2 rockets (many only partly assembled) together with many of the leading German designers. A number of V2s, and some of the men who designed them, were taken to the United States to assist in the post war American rocket programme. The captured rockets were assembled and launched primarily for the purposes of military training and assessment, but since the V2s were capable of reaching altitudes much greater than balloons, the US Army Ordnance Department announced the opportunity to fit some of the rockets with scientific payloads. The first such launch, using a V2 instrumented for studies of the upper atmosphere, was from the White Sands missile range, New Mexico on 10 May 1946. The potential of the V2 for astronomical research was also evident; the rocket could lift a 1000 kg payload to 160 km, providing about 5 minutes of observing time above most of the ultraviolet absorption in the ozone layer.

The first attempt to make rocket borne astronomical observations, using an ultraviolet spectrograph fitted in the nose of a V2 launched in June 1946, was not a success. The instrument was carried to a height of almost 120 km before the V2 fell back to Earth, ending its flight in an explosion which produced a crater 10 metres across. It is said that the remains of the spectrograph exhumed from the desert sand after weeks of digging would just about fill a wastepaper basket. About the same time another astronomical experiment was attempted by scientists from the US Naval Research Laboratory (NRL) who used V2s to carry a simple device aimed at detecting X-rays from the Sun. A piece of photographic film, shielded by a sheet of beryllium foil impervious to visible and ultraviolet light, was lifted above the absorbing layers of the atmosphere. The idea was that solar X-rays, if they existed, should be able to penetrate the foil and fog the film. On 6 August 1946 T. R. Burnight reported the detection of solar X-rays by this method, and this is widely held to be the first such detection, although it has been suggested that the film was blackened, not by X-rays, but by impact pressure as the rocket returned abruptly to Earth (see the review by Pounds, *Q. J. R. A.S* **27** p 436 (1986)).

In any event, 1946 was a crucial year for space astronomy because on 10 October a V2, carrying a small spectrograph built by the NRL, obtained a solar spectrum which stretched much further into the ultraviolet than any previous observation. The NRL team had learned from their previous experiences; this time the spectrograph was fitted in an armoured package mounted at the tail of the rocket and the vital film was recovered successfully. Other rockets were launched to continue and extend these research programmes. An unambiguous detection of solar X-rays was made by an NRL rocket carrying a Geiger counter, and, over the next few months, more and more detailed ultraviolet spectra of the Sun were obtained.

During the period 1946 to 1951 sixty-nine V2 rockets were launched, and, according to James Van Allen, about forty of these could be counted as successful. Meanwhile, rocket development proceeded apace. In 1949 the United States launched its first Viking sounding rocket, and by 1954 a Viking had soared to an altitude of over 250 km and laid the ground for the development of the Vanguard satellite programme. Another important sounding rocket of the same period was the Aerobee, developed by the Aerojet Engineering Corporation and the Douglas Aircraft Company under the direction of Van Allen, and first launched in 1947. The Aerobee evolved steadily into a vehicle capable of lifting a 270 kg payload to an altitude of over 300 km, and continued in service until 1985 by which time over one thousand had been launched, many on astronomical missions.

Outside the USA, Canada developed the Black Brant series which, like the Aerobee, has undergone many developments since its first flight in 1959. The latest version, the Black Brant XII, is capable of reaching altitudes of 1600 km. The UK also produced a successful high altitude sounding rocket, the solid fuelled Skylark. The first Skylark was launched in 1957 as part of the International Geophysical Year, and, like its competitors has been steadily improved over the years. The original single stage Skylark was spin stabilised, rotating about its long axis during flight, but later versions contained a stabilisation system able to lock onto a star and provide accurate pointing for experiments. The Skylark 12, a three stage version, is able to lift a payload of 100 kg to 1000 km. Skylarks were used in numerous astronomy missions, but are now mostly used for materials research.

The use of sounding rockets for astronomical research has continued into the era of satellites, since, like balloons, rockets offer certain advantages over satellite missions. These are summarised below.

The launch of a Skylark sounding rocket (BAe)

(1) *Economy*

Although short, sounding rocket flights are much cheaper than satellite experiments, so offer an opportunity to investigate new ideas relatively cheaply. An example is the application of coded mask imaging in high energy astronomy (section 2.5.4). This revolutionary technique was tested on a Skylark rocket flight in 1975, and, having proved its usefulness, was developed into a major Spacelab experiment which flew on the Space Shuttle in 1985. The low cost of sounding rocket flights also means that scientific groups may get the opportunity to fly several missions and to recover and improve their instruments and operating techniques between each one. Satellite experimenters seldom get such an opportunity.

(2) *Reliability*

Early space launch vehicles were experimental, and the failure rate was high. Many expensive and sometimes irreplaceable satellites never even reached space (e.g. see section 4.4.3) because of launch vehicle failures. By comparison simple, often solid fuelled, sounding rockets are fairly reliable, and, if one rocket fails, another payload can usually be assembled and launched after only a short delay.

(3) Flexibility

Because they are simple, sounding rocket flights can be made at short notice. This is obviously an advantage for research into transient phenomena like the Aurora, but it also makes it possible to attempt important measurements of unusual astronomical events with a high probability of success. For example on 7 March 1970, 25 rockets of various types were launched from the NASA Wallops Island facility in Virginia to observe conditions before, during, and after a total solar eclipse. No fewer than 11 rockets were launched within a six minute period during the period of totality. Coordinating a number of satellites to make similar observations would be difficult if not impossible, even if the correct types of instruments were in orbit and operating at the time.

(4) Short lead times

In general a satellite mission takes many years to come to fruition; a typical 1980s astronomy satellite can take ten years or more to get from drawing board to orbit. In comparison a sounding rocket experiment can be designed, produced, flown, and the results analysed in perhaps two years. This is a much more suitable timescale for many small projects, especially those involving students working for higher degrees.

(5) Simplicity

Because of their low cost and short flight times, payloads for sounding rockets do not need to be as reliable or as thoroughly tested as those for satellites. It is often cheaper to build a simple instrument and try again if it fails, than to build in the complex redundant systems required to achieve the high reliability required of satellite experiments.

Sounding rockets made many of the important breakthroughs in space astronomy. The first X-rays from the Sun, the first X-ray star, and the first ultraviolet spectra of stars beyond the solar system were all observed during rocket flights. The steady advance of space technology has, however, diminished the role of sounding rockets. As launchers became reliable, and the Space Shuttle promised cheap and regular access to space, sounding rocket programmes were cut back. Some nations, including the UK, abandoned them completely.

However, as in the case of balloons, the advantages of maintaining a sounding rocket programme were demonstrated in 1987 with observations of the supernova in the Large Magellanic Cloud. The supernova, the first example visible to the naked eye for 400 years, was of enormous interest to astronomers, and in an attempt to maximise the scientific return from this event several sounding rocket flights were organised to supplement studies made from the ground and from orbit. The first of these was by a German group using a Skylark rocket carrying an X-ray telescope. Despite a failure of the rocket's star tracker at a critical moment, a successful observation was made as ground controllers took command and guided the rocket manually, using information from a television camera on board the payload.

1.5 ASTRONOMY FROM SATELLITES

The idea of placing telescope in orbit predates the space age by at least half a century. In 1923 the German rocket pioneer Hermann Oberth wrote about the advantages which a space telescope would have over a similar instrument on the ground, and lunar observatories were a favourite setting for science fiction novels of the 1950s. Despite this, when the first artificial satellites were launched, astronomers were not at the front of the race to place their instruments in orbit. Many of the first scientific satellites were aimed at studying the properties of space in the vicinity of our planet, not for observing distant stars and galaxies, and it was some years before the first true astronomical satellites were launched. Some of the reasons for this delay were technical ones, but others were probably similar to the reasons why, even thirty years later, some astronomers are reluctant to take part in space projects. To understand this, the advantages, and disadvantages, of trying to do astronomy from a satellite must be considered.

The advantages of taking an observatory above the atmosphere are obvious. The absence of absorption opens up the entire electromagnetic spectrum to study, and, unlike a rocket flight which lasts only minutes, satellites can operate for many years. These two factors alone can provide enormous advantages for astronomy, but satellites have other characteristics which can present considerable professional risks for the individuals concerned with their development.

For example, the very high cost of space experiments makes it unlikely that one astronomer, or even one research institute, will be allocated enough money to develop a satellite for their own use†. More often, groups are allocated funds to develop a single instrument which must then share space on a satellite with other experiments. This, together with the involvement of government agencies and large aerospace companies, means that the astronomers must spend considerable time in complex and often tedious negotiations to ensure that all the experiments can be accommodated on the spacecraft and that sufficient funds are available to develop the satellite and provide the required ground support.

Worse still, factors unconnected with the scientific objectives of a mission may mean that political and financial matters eventually determine the success or failure of a project, and thus affect the careers of the astronomers involved. An example of this is the fate of some of the experiments scheduled to fly in the High Energy Astronomical Observatory (HEAO) programme (see section 2.4.6). In 1973 both the HEAO programme and the Viking Mars project were encountering financial problems; and for political reasons NASA, the agency responsible for both programmes, was unable to seek more money from the American Congress. Funds for one of the projects had to be cut off, and, since Viking was a prestigious mission to search for life on Mars, the axe fell on the less glamorous HEAOs. After a period of intense lobbying a reduced HEAO programme was eventually resurrected, but in the process a number of the planned experiments were dropped. For the astronomers involved in the abandoned experiments years of work were wasted, and even for those who remained in the project, eighteen months had been lost in the political wrangling needed to get the project restarted. The ultimate result is that, even for those experiments which survive the financial, political, and technical obstacles along the way, a very long time can elapse before the satellite is ready to face the final hurdle and be placed into orbit.

The launch is probably the most agonising part of a satellite project since it is one over which the scientists have virtually no control. Launches can be delayed for a

---

†There are a few exceptions to this general rule, for example the NASA Small Astronomy Satellites.

variety of technical and political reasons, and even when the satellite finally lifts off there is the ever present possibility that, however reliable this type of launcher has proved in the past, a random failure may still claim this particular one. Once in orbit, things can still go wrong, solar panels may jam, protective covers may fail to eject, electrical discharges may destroy the sensitive instruments the first time they are switched on, and a host of other potential catastrophes can occur. Finally, even if everything goes perfectly, there is no guarantee that the satellite will produce sufficient new results to justify the enormous cost of its construction.

It is this uncertainty about the probability of success that has set the pace as far as the development of astronomical satellites is concerned. Governments, through their scientific committees, are reluctant to commit money to expensive projects unless there is at least a reasonable likelihood of success, and this has led to a step by step approach to space astronomy which is based on keeping the early stages as inexpensive as possible. The first step, and the one which requires the most vision, is for a small group of astronomers to be convinced that it is worth exploring the sky at certain wavelengths, or with a certain type of instrument. Next, some evidence to support this idea must be obtained, preferably as cheaply as possible. This is usually done by balloon or rocket flights aimed at objects which theory predicts will radiate strongly at the wavelengths of interest. With luck the observations will confirm the predictions and will also detect some totally unexpected phenomena.

If interesting sources are discovered by this almost random probing of the sky, and historically they always have been, then the next step is to carry out a survey (usually by a small satellite) to locate as many new sources as possible. Since a survey mission has limited capability for follow up measurements, a successful survey will lead to a demand for larger and more sophisticated satellites to make detailed studies of the most interesting sources. These 'observatory' satellites, which mark the next step in the development process, are often operated in a similar way to ground based telescopes; astronomers request observing time to study objects in which they have a particular interest and may participate directly in the process of making the observations.

After one or more successful observatory missions the next stage is the development of 'facility' class satellites, very large spacecraft which can be serviced in orbit. The basic spacecraft is designed to remain in operation for many years, but is constructed so that new instruments can be fitted by visiting astronauts to ensure that the orbiting facility remains at the forefront of astronomical research. Various facility class missions are already under development or in an advanced stage of preparation, and will be described in some detail in later chapters. Occasional, low cost, missions designed to tackle specific scientific questions may also be undertaken in this phase. Beyond these missions the path is less clear; many advanced concepts are being studied and some will be described in Chapter 8, but only time will tell how many of them ever come to fruition.

1.6 DEVELOPING AN ASTRONOMICAL SATELLITE

Despite the frustrations and professional risks, the scientific rewards from a successful satellite mission are huge, and there is not usually a shortage of astronomers willing to participate. This is not the place to enquire what makes some scientists prepared to invest so much of their time in space missions (but see W. & K. Tucker's *The Cosmic*

Enquirers); but it is worth considering what goes on in the long years before a satellite is launched. An excellent description of the early years of the Orbiting Solar Observatory programme, a typical scientific project of the 1960s, is given in A. Bester's *The life and death of a satellite* (see Bibliography), but some aspects of developing a satellite mission have changed considerably since then.

The first step, however, remains the same; someone must think of a good idea which is both technically feasible and scientifically worthwhile. This germ of an idea must then be developed into a proposal likely to command support from the rest of the scientific community and thus attract sufficient funding to make it a reality. This is usually done by the originators of the idea writing papers, giving lectures, and meeting other scientists to see if a consensus of objectives will emerge. If a general strategy can be agreed, then the next step is to prepare a detailed proposal for further consideration. The best time to do this varies, but space agencies like NASA and the European Space Agency (ESA) occasionally issue a request for proposals, essentially asking scientists to suggest what the objectives of future missions should be. The level of interest at this stage is usually high, as can be judged by what happened in 1983 when ESA asked for proposals for future European space science missions; seventy-seven detailed replies, describing sixty-five possible missions, were received.

What happens next depends on the workings of the space agency concerned, but in general once a number of suitable proposals have been received they are evaluated by various committees until a small number have been identified as the most promising candidates for further consideration. More detailed studies are then made to determine the feasibility of the proposals. This process is often referred to as Phase A, although such labels are rather vague and may mean different things in different projects. The expressions 'pre-Phase A' and 'assessment study' are also sometimes used to describe early work. The results of the Phase A studies are then used to decide which mission, or missions, will be chosen for further development. Next come further studies (Phase B) during which the engineering and financial aspects, as well as the scientific goals, of the proposal are evaluated in more detail until the mission is sufficiently well understood that contracts for the industrial development of the satellite and its payload can be placed. With the award of contracts for construction of the satellite the final design work can begin, and, eventually, the manufacturing and testing (Phase C/D) commences. The details of the development process vary considerably from project to project, depending on factors such as cost and schedule pressure, but what follows in this section can be considered as a general guide. In practice political and financial pressures may intervene to modify the development programme, and various problems with the spacecraft, or with its scientific payload, will considerably disrupt the best laid plans.

After the basic design of the satellite is complete, the work progresses through the construction and testing of a series of development models. Lessons learned during the manufacture and testing of these prototypes are fed back into the final design to rectify any problems found along the way. As an example, a development programme might require the construction and testing of an electrical model and a structural model to verify the soundness of the basic design, followed by the production of a flight model, the one destined to fly in space.

In such a three model programme the electrical model would be used to assist in the development of electromechanical subsystems such as motors, detectors, computers,

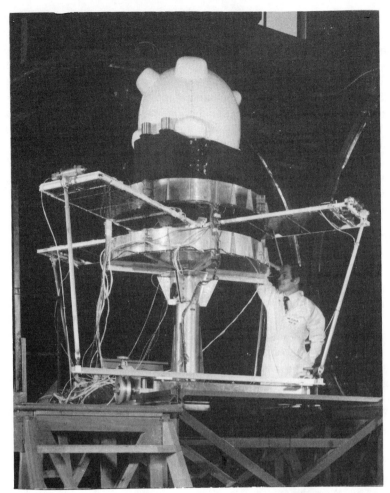

The Ariel 6 satellite is prepared for tests in a vacuum chamber. (SERC).

and tape recorders, and to test the wiring harnesses needed to connect them all together. The prototype electrical systems are fitted to a structure which is representative of the final design, but which has not necessarily been fully developed. A key function of the electrical model is to test for interference between different electrical systems; for example, does switching on a motor cause a surge in the power supply which in turn causes the onboard computer to fail? This type of investigation in known as Electro-Magnetic Compatibility (EMC) testing and is also used to confirm that the satellite will not produce any stray electrical signals which might interfere with the launch vehicle or disrupt the scientific instruments. The electrical model can also be used to develop the procedures which will be used to check the satellite before launch and to operate it once it is in orbit.

The structural model is used to verify the mechanical properties of the design. Since, during testing, it will be subjected to levels of vibration beyond those likely to be encountered during an actual launch, it usually contains a minimum of working

Developing an astronomical satellite

electronics. The place of the computers and sensitive electronic components is taken by less expensive 'mass dummies' with the same mechanical properties as the real equipment. The use of mass dummies also means that development of the electronic subsystems can continue without interruption during the mechanical test programme. Probably the most critical stage of testing is the vibration test when the structural model is placed on a special test stand and shaken violently to confirm that the structure will survive the vibrations encountered during launch. During vibration testing the structural model is fitted with accelerometers which record the effects of the test on various parts of the satellite.

The structural model may also be used for thermal testing, in which case the mass dummies must be fitted with heaters to simulate the energy dissipated by the electronics in the real satellite. Thermal testing takes place in vacuum chambers which can be cooled to very low temperatures (e.g. to 77 K by pumping liquid nitrogen through

The Scout rocket was used for many small astronomical missions because of its low cost. This is the launcher for the Ariel 6 satellite described in Chapter 2. (SERC Photo courtesy Marconi Space Systems).

special shrouds surrounding the satellite) and may be fitted with powerful spotlights to shine artificial sunlight into the chamber. In this way it is possible to carry out a full simulation of conditions in orbit and hence verify that the thermal design of the satellite is correct.

The flight model is the final version. Only when the overall design has been proved by tests of the electrical and structural models, and approved by independent experts in a series of design reviews, does production of the flight model proceed. Once assembled, the flight model undergoes a series of 'qualification tests' to verify that it is ready for launch, and it is then shipped to the launch site for fitting to the vehicle which will carry it into space. This last step is usually known as 'integration'.

The structural and electrical models are usually kept in reserve after the flight model is complete. They may be used as training aids for ground staff, from crane drivers to mission controllers, or for troubleshooting if the satellite develops problems once in orbit. In the event of a launch failure, they can also be used to provide parts for the production of a replacement satellite.

The Atlas–Centaur rocket, here launching the ill-fated OAO-B, is the largest NASA launcher used for space astronomy payloads. (NASA).

Naturally, not all missions are developed by means of a three model programme. If the satellite is large, the cost of producing three sets of near identical hardware may be excessive, and the electrical and thermal models may be combined into a single 'engineering model'. Alternatively, only a single model may be built. This 'protoflight' version is used for all the prelaunch testing and is then refurbished to incorporate any modifications required after the results of the test programme have been analysed. The protoflight philosophy may result in considerable cost savings, but leaves nothing available from which to assemble a reserve satellite in the event of a launch failure. Additional disadvantages of a single model programme include the risk of overtesting critical components, which may result in premature failure after launch, and the ever present danger of having to begin a complicated and expensive redesign part way through the test programme if major problems arise.

1.7 THE CHOICE OF ORBIT

The details of project management techniques and of satellite design and construction can be found elsewhere, but a few of the factors affecting the choice of orbit for an astronomical satellite will be mentioned. This is a most important decision since the orbit will determine the size, and cost, of the vehicle which must be used to launch the satellite (if the size of the launcher is already fixed by budgetary or other constraints, the chosen orbit will determine the mass of the satellite). The decision is seldom clear cut since as well as the basic mission objectives, factors such as ease of communications, protection of the satellite from background radiation, sky coverage, and mission lifetime must be carefully considered. Ultimately, engineering and scientific compromises are always necessary if a practical solution to often conflicting requirements is to be found.

1.7.1 Communications

Satellites must both receive instructions from ground controllers and return their data to Earth at regular intervals if they are to carry out their mission successfully. This makes ease of communication a major consideration in orbit selection because a satellite in a low orbit is in view of an individual ground station for only a few minutes at a time. If the satellite orbits close to the Equator it may be contacted for a few minutes every orbit (i.e. about once every ninety minutes or so), but in an orbit which passes over the poles, contacts with an individual ground station may be limited to a few short intervals every 24 hours. In principle, continuous contact can be achieved by having ground stations spread around the world, but in practice this is prohibitively expensive to maintain on a regular basis (although it was done during the early days of manned space exploration). Since scientific satellites are often placed in low orbits for other reasons (e.g. cost) the usual solution is to fit the satellite with an onboard computer which can be programmed with a series of instructions and to provide a tape recorder on which data can be stored. A satellite thus equipped can transmit all its recorded data, and receive a new set of instructions, during each brief period of ground contact. It can then operate automatically until the next communication session. A satellite in this class might have about 64 kilobytes of memory and tape recorders capable of storing in the region of 450 megabits of scientific and 'housekeeping' data. Included in the

A typical ground station antenna used to communicate with astronomical satellites. This is the antenna at the Rutherford-Appleton Laboratory near Oxford, UK, which was used for the IRAS project in 1983. (SERC).

programs stored in memory (usually in ROM) the satellite must carry a set of emergency instructions to allow it to place itself in a safe condition if a malfunction occurs when it is out of ground contact.

A low orbit, in which a satellite spends most of its time operating automatically, is acceptable for a survey mission, since most of the satellite's time is spent scanning the sky and storing data for subsequent analysis, but it is not so desirable for an observatory mission. To obtain maximum scientific return from an astronomical observatory, in space or on the ground, it is usual to allow the astronomers to play an active part in the observations and to allow them access to the data as it is being collected. This allows the quality of the data to be evaluated immediately, and makes it possible to modify the observing plan if necessary. Obviously this cannot be done in the case of a satellite which is out of contact for most of the time.

Future observatory satellites operating in low orbit may maintain more or less continuous contact with ground controllers by using a satellite in a higher orbit as a relay station. This is the function of the NASA Tracking and Data Relay Satellite System (TDRSS) which will be used to communicate with the Hubble Space Telescope (see section 5.1). Although TDRSS data links will be important in the future, historically they have not been a feature of space astronomy missions. To date, the only way to ensure virtually uninterrupted communications with an observatory satellite has been to place it in a high orbit which is in view of a single ground station for long periods. One such orbit is the geostationary ring, 36 000 km above the Earth, used by communications satellites. At this height a satellite revolves around the Earth every 24 hours, and so appears to hover over a single spot all the time, making round the clock coverage from a single ground station possible. Unfortunately, geostationary orbit is difficult and expensive to reach, and, since it offers few advantages other than ease of communication, it has not been used for astronomical missions in the past (but see section 5.2 on Hipparcos). The usual alternative is a highly elliptical orbit, arranged so that as the satellite climbs to its apogee it remains in sight of a ground station for a considerable period, with the precise length of the contact depending on the satellite's orbit. A satellite in an elliptical polar orbit will seem to hover above one or other pole for much of the time, allowing long contacts with a single station in the correct hemisphere. For example, a satellite in a near polar orbit ranging from 500 to 200 000 km will be accessible from a single ground station for all but a few hours of each 99 hour orbit. A similar orbit close to the Equator will require two ground stations at different longitudes to maintain constant coverage, since the rotation of the Earth will sweep each station out of sight of the satellite for about 12 hours in every 24.

1.7.2 Background radiation

The altitude at which an astronomical satellite can operate, and the inclination of its orbit with respect to the Equator, will also be influenced by the need to reduce background radiation to an acceptable level. This background may be considered to have two components: stray photons produced in the tenuous upper atmosphere which can swamp the light from faint astronomical sources, and charged particles which penetrate the sensitive instruments and cause false detections.

The photon background is particularly intense in the far ultraviolet. It is dominated by intense emission of certain wavelengths due to scattering of solar radiation by ions and neutral atoms in the upper atmosphere, and is referred to as geocoronal radiation. The most important wavelengths at which emission occurs are 30.4 nm and 50.4 nm (due to doubly and singly ionised helium respectively), 1216 and 1025 nm (hydrogen emission lines), and 1304 and 1356 nm (emission lines of singly ionised oxygen). The amount of emission varies with viewing angle and is considerably reduced in the Earth's shadow.

In other wavelength regions it is charged particles, trapped in the Earth's radiation belts, which produce the unwanted background detections. The belts, named after American physicist James Van Allen, consist mostly of protons and electrons spiralling around in the Earth's magnetic field. The inner belt, which contains protons with energies of around 50 MeV and electrons with energies of greater than 30 MeV, lies between about 1000 and 5000 km altitude. In some places, notably the so called South Atlantic Anomaly, a considerable concentration of particles is also found at lower altitudes. The outer belt, between about 15 000 and 25 000 km, contains less energetic

The low density hydrogen gas in the Earth's geocorona revealed by the Apollo 16 ultraviolet camera. Note that the emission is far more intense on the sunlit side of the planet (left). A faint arc of emission (lower right centre) marks the site of an aurora near the South Pole. Geocoronal radiation may present a major problem for satellites carrying certain types of ultraviolet experiments. (NRL/NASA).

particles and varies in intensity. Particles trapped in the Van Allen belts have a serious effect on many of the electrical components on board a satellite, and sensitive astronomical detectors are particularly vulnerable. Charged particles are capable of producing false signals (i.e. background noise) in most types of gamma ray, X-ray, and infrared detectors.

Little can be done about the geocoronal radiation except to try to suppress it with suitable filters or to observe when the satellite is looking along the direction of the Earth's shadow, but clearly astronomy satellites must operate in orbits which avoid the Van Allen belts as far as possible. One option is to use a low orbit which remains below the inner belt, although this causes communication difficulties and may expose the satellite to excessive atmospheric drag and shorten its lifetime; another is to use a very high orbit above the outer belt, but this demands a powerful and expensive launch vehicle. A compromise is to use a highly elliptical orbit in which the satellite spends most of its time above the radiation belts, and to suspend scientific observations during the time spent passing through the belts. With careful planning it is possible to arrange that the passages through the Van Allen belts coincide with periods when the satellite is out of sight of its ground station and operations would normally be suspended because of communications difficulties. In this way the time when the satellite is unable to observe can be minimised.

1.7.3 Sky coverage

The ease with which a satellite can observe a given region of the sky is also an important factor in the choice of orbit. For a survey satellite any orbit which enables the whole sky to be covered in a reasonable time will suffice, but an observatory mission requires a capability to point at many different targets in an almost arbitrary order and to observe them for comparatively long periods.

Survey missions usually use a low orbit (to reduce launch costs) in which the satellite's instruments sweep out a great circle from pole to pole every orbit. This can be done by orienting the spacecraft so that one face is always pointing towards the Sun, and having the instruments observing at right angles to this line, known as the Sun vector. If the spacecraft is rotated once per orbit about the Sun vector, its detectors will sweep out a circle on the sky each time it travels around the Earth. If the Earth were perfectly spherical, the orbit would remain fixed in space and the satellite would sweep out the same strip of sky over and over again, but fortunately this is not the case. The Earth is not a perfect sphere because, amongst other things, the rotation of the planet causes it to bulge slightly at the Equator. The bulge is only about 0.33% of the Earth's radius, but this is enough to exert a non-uniform gravitational force on a satellite. The effect of this force is able to change the orbit of the satellite slowly, an effect known as precession. The theory of orbits can be found in other texts, so we will merely note that

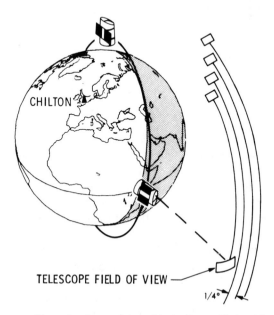

Sky survey from a polar orbit. As the satellite's orbit changes because of the uneven tug by the Earth's equatorial bulge, the satellite sweeps out overlapping scans of the sky. The example is the Infrared Astronomical Satellite IRAS, described in Chapter 7. (NASA).

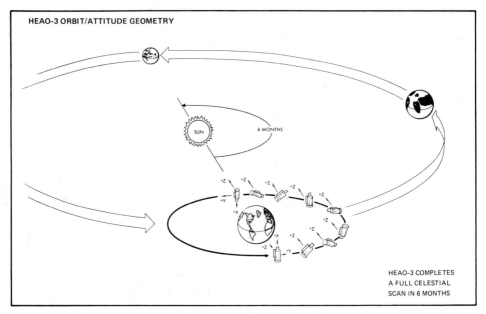

Sky survey from an equatorial orbit. By rotating about the Sun vector the satellite sweeps out scans of the sky. The example is the HEAO-3 satellite described in Chapter 3. (NASA).

the satellite's orbit rotates around the Earth's axis while keeping the same inclination relative to the equator, and the satellite's orbit also rotates in its own plane.

With careful planning it is possible to use these effects to change the plane of a satellite's orbit throughout its lifetime and thus allow it to survey the entire sky. One method is to choose a polar orbit that rotates by about 1 degree per day, thus keeping the same side of the satellite facing the Sun as the satellite follows the Earth on its annual journey around the Sun. Such an orbit is said to be Sun synchronous, and an instrument with a moderate field of view which points at right angles to the Sun vector will cover the entire sky in six months. The coverage will, however, not be uniform. The instrument will sweep across the ecliptic poles on every orbit, but it will sample only a small portion of the ecliptic equator each day. The effect of this non-uniform coverage must be allowed for when analysing the results of any such survey.

Some survey missions use a proportion of the time which would otherwise be spent repeatedly scanning the poles to carry out routine engineering tests that disrupt the data flow, or use the time to make short pointed observations of other targets. Mission planners may also pick an object near the pole as a calibration source and observe it on every orbit to monitor any changes in the sensitivity of their instruments as the mission progresses.

Although low orbits are well suited to survey missions they are not so desirable for satellite observatories, since a low orbit severely restricts the freedom of a satellite to point at objects of interest. This is because of the considerable angular size of the Earth when seen from low orbit; from 500 km altitude the Earth subtends an angle of 106°, blocking out a sizeable fraction of the sky. When allowance is made for engineering factors, such as the need to protect sensitive star trackers from being blinded by the bright Earth, restricting the amount of stray light falling onto the astronomical instruments, the need to point antennae, etc., more than half the sky is unavailable for study at any given moment.

The short orbital period of a satellite close to the Earth also means that the portion of the sky available for study changes very quickly. A satellite in a low equatorial orbit will be unable to make uninterrupted observations of a target near the celestial equator because the object will disappear behind the Earth at regular intervals (the same effect means that the sunlight to the solar panels is periodically cut off, and rechargeable batteries must be used). The only way to study an equatorial target for more than a few tens of minutes is to make a series of short pointings and then add the results together during the data processing. Continually repointing the satellite in this way complicates the observation planning, may shorten the lifetime of the mission by depleting the supply of attitude control gas, and makes data analysis more difficult. Similar observational problems would be encountered in a low polar orbit, only in this case it is the regions around the celestial poles which cannot be studied for long periods.

These difficulties do not mean that it is impossible to make detailed observations of selected objects from a low orbit; they just make it more difficult. There may be reasons why the satellite must operate close to Earth, for example to allow it to be reached by Space Shuttle servicing missions or because the available launch vehicle is not powerful enough to reach a greater altitude. In this case the mission must either use a data relay satellite or carry out preprogrammed observations under the control of its onboard computer. In the latter case direct control during the observations is not possible, although the satellite may still be referred to as an observatory.

Observatory missions are better suited to high orbits from which the Earth appears much smaller (our planet subtends an angle of only 7° when seen from a satellite 100 000 km away), allowing long uninterrupted pointings over almost the entire sky.

1.7.4 The effects of residual atmosphere

Another factor to be considered when planning a satellite mission are the effects of the residual atmosphere at the spacecraft's operating altitude. The most obvious manifestation is atmospheric drag which, even at altitudes of over 1000 km, is sufficient to produce measurable changes of a satellite's orbit. Each time the satellite passes its perigee (its closest point to the Earth) atmospheric drag slows it slightly, and the satellite cannot climb all the way back to its original apogee (its farthest point from the Earth). The orbit gradually becomes more circular, and eventually so much energy is lost during each perigee passage that the satellite rapidly slips into the denser atmosphere and is destroyed during re-entry. The likely lifetime of a satellite before re-entry can be calculated, although changes in the density of the upper atmosphere caused by the activity of the Sun make this an uncertain business, and this must be taken into account by mission planners. There is no point in providing sufficient consumables (attitude control gas, etc.) for a 10 year mission if the satellite will re-enter after 2 years in orbit.

The thin upper atmosphere also presents a more subtle threat to delicate astronomical instruments in the form of contamination of reflecting surfaces and filters. One possibility is chemical reactions occurring on optical surfaces, another is the condensation of gases onto components of ultracold infrared telescopes. Direct chemical effects are believed to have been responsible for a dramatic fall in the reflectivity of the mirrors used in an X-ray telescope carried on the UK 6 satellite. The mirrors appeared to have degraded during or soon after launch, and this has been ascribed to the effect of traces of oxygen found at the satellite's orbital altitude on the gold coated mirrors. Chemical degradation of metallic filters planned for use in extreme ultraviolet instruments has also been reported, with experiments on the Space Shuttle indicating that residual atomic oxygen was responsible for the damage.

Condensation of gases is likely to occur on cold optical surfaces such as those used for infrared astronomy; even at 1000 km altitude there is sufficient atomic oxygen to condense on cryogenically cooled surfaces. Individual oxygen atoms deposited on a surface colder than 20 K will eventually combine to form molecules, and, once formed, the molecular oxygen will remain almost indefinitely. Calculations have shown that a layer of frozen oxygen 1 μm thick could accumulate on an unshielded cold mirror in less than a month unless special precautions are taken. One method to protect cooled satellites against this kind of contamination is to limit the pointing directions so that the aperture of the telescope is never allowed to point 'into wind'. This prevents the telescope scooping up gas atoms as it circles the Earth.

One other effect of the atmosphere on a satellite in low orbit is worthy of consideration. Observations from the Space Shuttle have shown that a bright glow builds up around the craft once it is in orbit. This is attributed to chemical reactions involving atmospheric atoms which occur on the surface of the Shuttle. Stray light from these reactions may have been responsible for fogging photographs taken by an ultraviolet telescope on the Spacelab 1 mission (section 4.9). If confirmed, this glow may limit the usefulness of satellites in low Earth orbit for certain types of ultraviolet studies.

1.8 SOME OTHER DESIGN CONSIDERATIONS

1.8.1 Attitude control systems

The choice of attitude control system for an astronomical satellite is determined by both engineering and scientific factors. Some early survey satellites were spin stabilised, using gyroscopic effects to maintain a constant spin axis while their instruments scanned around the sky. The pointing directions of these satellites were adjusted by either gas jets or magnetorquers (electromagnets on the satellite which react with the Earth's magnetic field to manoeuvre the spacecraft). Gas jets can produce rapid manoeuvres if required, but are limited by the amount of propellant carried, while magnetorquers are reliable and can operate for as long as the satellite's power supply remains operational, but cannot reposition the satellite rapidly. Spacecraft which rely on magnetorquers always run the risk that if the solar panels lose their lock on the Sun, the spacecraft batteries may run down before the magnetorquers can turn the satellite enough to restore the power supply. This fate nearly befell the Solar Maximum Mission spacecraft in 1984 during attempts by astronauts to repair the ailing satellite.

Satellites which need to be pointed with considerable accuracy often use what are known as reaction wheels. By applying a force between the satellite structure and a spinning wheel, angular momentum can be exchanged between the wheel and the structure, causing the satellite to rotate. By using three mutually perpendicular wheels it is possible to point a satellite in any direction. Satellites with reaction wheels must,

Some other design considerations

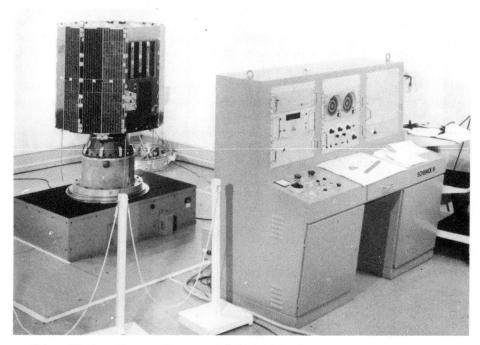

Spin stabilised satellites must be accurately balanced before launch. Here the Ariel 5 satellite, described in Chapter 2, undergoes tests on a spin table. (SERC).

however, carry an additional attitude control system, either gas jets or magnetorquers, to alter the angular momentum of the wheels if they approach their operating limits. This is done by using the second system to start the satellite moving, then using the reaction wheels to stop it again. In this way the wheels can be speeded up or slowed down as required. The choice of the second system may be determined by cost, expected mission lifetime, or the need to avoid contaminating the satellite with gas expelled from thrusters. Magnetorquers are most frequently used, especially for satellites expected to have a long operational lifetime. The relative capability of reaction wheels and magnetorquers can be seen by comparing the attitude control actuators on the IRAS satellite. The maximum torque from the reaction wheels was almost 100 times greater than that from the magnetorquers used to unload them when required.

1.8.2 Cooling of scientific instruments

Maintaining all the components of a satellite at the correct temperature is always a problem since at any moment a satellite may be receiving heat in the form of sunlight or earthlight, be generating heat from its electronics, and yet be radiating energy away from parts of the structure which are pointing directly at cold space. To add to the complications the situation may change rapidly when the satellite goes behind the Earth and sunlight is cut off for a time, or when the satellite moves to point in another direction. Designers of astronomical satellites may also have to accommodate equipment which must be maintained at low temperatures to operate efficiently;

examples of such instruments are solid state gamma ray detectors and most kinds of infrared instruments.

The simplest and most reliable method of cooling equipment is to arrange that it is mounted in such a way that it always faces space and never receives any direct sunlight. In this situation the equipment will radiate away considerable quantities of energy, and, depending on the detailed design, may stabilise at quite a low temperature. This type of passive cooling is often adequate for electronic detectors which need to be kept a few tens of degrees cooler than other parts of a satellite.

If an instrument cannot be cooled sufficiently by simple radiation, some form of active cooling will be required. In an open loop system the instrument is surrounded, or at least kept in contact with, a reservoir of cold material. Heat from the instrument leaks into the coolant and boils it away. The coolant chosen depends on the temperatures at which the instrument needs to be maintained, and could be solid (e.g. frozen ammonia or frozen methane) for temperatures around 100 K or liquid (e.g. liquid helium or liquid neon) for very low temperatures. Such a system is simple and

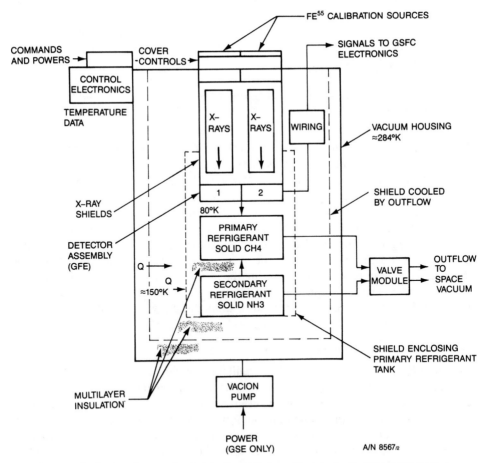

An example of an open loop cooling system; the subliming refrigerator for the HEAO-2 Solid State Spectrometer. (Ball Aerospace Systems).

reliable but limits the lifetime of the instrument; once all the coolant has boiled away the instrument warms up and is useless.

Closed loop systems do not rely on boiling away a limited supply of coolant but circulate a working fluid (or gas) around the instrument. The fluid absorbs energy from the instrument and carries it away, cooling the instrument in the process. The fluid is then passed through heat exchangers which remove the excess energy and then radiate it away into space. Closed loop cooling systems do not have any predetermined limits on their operating lifetimes, but are often heavier and less reliable than open loop systems and may require considerable amounts of electrical power to drive them.

1.8.3 Design flexibility

During the first two decades of the space age individual scientific missions were relatively simple and were developed and launched over a period of a few years. In recent times this situation has changed as satellites have grown ever more complex and expensive and as the use of space for commercial purposes has increased, placing greater demands on the available launch vehicles. Compounding the problems faced by scientific missions has been the decision by the United States, seen to have been mistaken after the loss of the Challenger and her crew of seven astronauts, to abandon the use of expendable rockets in favour of the re-usable Space Shuttle.

As a consequence of these changes several astronomical missions, for example the Hubble Space Telescope (HST) and ROSAT, have faced long launch delays which have led to problems that were not anticipated when the spacecraft were designed. The Hubble Space Telescope (see section 5.1) was already several years behind its originally planned launch date when the Challenger exploded in January 1986. The long hiatus in the Space Shuttle programme caused by this disaster led to a further delay in the launch of the HST which has had a potentially serious impact on the mission. Firstly, the prolonged period of storage has led to fears about the deterioration of several important elements of the satellite, including the possibility that the main mirror may become contaminated. Since the mirror is deep within the satellite and cannot be reached without virtually dismantling the entire spacecraft, there is no real possibility of cleaning it before launch, and any degradation will have to be accepted. Secondly, advances in scientific instrumentation since the original payload was designed are in danger of rendering the instruments aboard the HST obsolete before the mission is launched. Second generation scientific instruments are already being prepared for eventual installation in the HST, and the validity of launching such an expensive mission without replacing the instruments first has already been questioned.

ROSAT (see section 2.6) was also designed for launch on the Space Shuttle and has also been significantly delayed. Fortunately, ROSAT is small enough to be launched by an expendable rocket, but the decision to transfer to a different launcher late in the development programme has forced last minute design changes to the satellite. The most obvious of these is the need to redesign the solar panels so that they can be folded to fit inside the nose shroud of a Delta rocket, but many other small changes are also required. As well as the cost of these changes, and the disruption to the test programme, much of the work already done (for instance most of the calculations on loads experienced during launch and the design of checkout procedures during deployment from the Shuttle) is now worthless and must be done again, causing further delays and increases in costs.

Similar problems have affected the Extreme Ultraviolet Explorer and two important deep space missions, the Galileo probe to Jupiter and an international mission, known as Ulysses, to probe the poles of the Sun. These have all been severely affected by repeated changes in the planned launch date and available launch vehicle, although the specific problems encountered will not be discussed here.

Although these particular problems cannot now be avoided, the lessons of the 1980s for future space science missions must be learned. Future missions must be designed with maximum flexibility so that they can survive long periods of storage without the danger of deterioration or obsolescence, and must be capable of being switched from one launcher to another with minimum cost and impact on the overall mission objectives. This will not be easy, but must be done.

1.9 SCIENTIFIC OPERATIONS

The scientific planning of the observations carried out by a satellite must take account of several factors. Most important of these is the need to obtain the best possible scientific return from the mission with the lowest technical risk, and enormous effort goes into preparing the observing plans. For a survey mission a few high priority regions of the sky may be identified for early study, although such observations must not detract from the overall objective of covering the whole sky in an unbiased way. The main consideration is to cover the entire sky as quickly and reliably as possible. For observatory missions the situation is more complicated because different astronomers will have different opinions of what constitutes an important observation.

The usual way in which the programme for an observatory mission is drawn up is to allow astronomers to define the observations they would like to make, and the reasons why they would like to make them, and to submit these proposals to a central coordinating committee. The committee then passes the applications to other astronomers with suitable experience who assess the requests in terms of their feasibility, originality, and their chance of producing important new results. The proposed observations are then given a grade, and the best proposals are added to the observing programme of the satellite. This process, which is similar to that used for allocating observing time on ground based telescopes, is called 'peer review'. At the end of each observation the data are given to the astronomer who proposed the idea, and this individual, or group of individuals, is allowed to analyse the data in private for a limited period, usually six months or a year. At the end of this period the data are placed in an open database, and anyone who wishes to can request a copy and analyse it himself. The reason for this process is to allow the originator of an idea a chance to publish his results first, but also to ensure that other astronomers have access to the data after a reasonable time.

In a perfect world all astronomers would have an equal chance to suggest observations for a space observatory, but in view of the enormous effort that it takes to organise a satellite project it is important to allow the individuals who have invested many years in the project an opportunity to benefit from their efforts. At the same time other astronomers must be allowed a fair chance to use the satellite since other, often equally worthwhile, projects may have been cancelled or delayed to provide funds for each mission which does take place. In the case of international projects it is also necessary to ensure that each nation receives a share of the observations which is in proportion to the capital which they have invested.

There are two common ways of resolving the questions of fair allocation of observing time. One is to form a 'science team' consisting of the people who have done most work in developing the satellite and to allow them some fraction, say 50%, of the observing time for themselves. The remainder of the observing time is allocated, by the peer review process, to 'guest observers'. In some cases, for example the NASA Small Astronomy Satellites, the science team may be allocated all the observing time. The science team, which may be composed of individuals from different institutes or from different countries according to some previously worked out formula, then define a core programme of the objects they most want to observe, and the guest observers must select their objects from what remains. An alternative is to select all the observations on scientific merit, but to allow the groups responsible for developing the satellite or its payload a number of the early observations to verify the performance of their instruments and to make certain high priority observations. This is sometimes known as 'guaranteed time'. If the satellite functions well, these groups inevitably get a quick look at what they consider to be the most interesting objects, and thus gain the benefit of a head start on their rivals.

Inevitably, ensuring fairness during this process is fraught with difficulty, and skilful political negotiating may be required to reach what everyone agrees is a reasonable arrangement. Nonetheless, in most cases the need to give the satellite developers a fair return on their investment of time is accepted, and suitable compromises are worked out. In the end, if the satellite fails, the division of data is irrelevant, and if it succeeds there is usually more than enough science to go round.

1.10 CONCLUSION

This chapter has reviewed the need for space observatories and has outlined the processes by which they come about. Much has been simplified or omitted, but it is hoped that the reader now has some idea of what goes on behind the scenes. The book now turns to a review of various astronomical experiments and the results which they have produced. Although the sequence of chapters generally deals with progressively lower energies, it begins with X-ray astronomy, for this, of all the disciplines, illustrates many of the concepts outlined above. Owing to limitations of space no attempt has been made to cover the large number of satellites devoted to purely solar astronomy.

2

X-ray astronomy

2.1 INTRODUCTION

It is convenient to regard X-rays as running from photons with an energy of about 0.1 keV (a wavelength of about 10 nm) to about 100 keV (about 0.01 nm). At high energies X-rays merge into gamma rays; at lower energies they join the extreme ultraviolet (XUV/EUV) region. Photons in the XUV region are also sometimes known as soft X-rays. To produce X-rays by thermal emission from a hot object (black body radiation) requires very high temperatures, typically in the range from 1–10 million degrees, and while astronomical sources this hot do exist, cosmic X-rays are more often produced by one of a number of non-thermal processes which occur in high temperature plasmas.

A plasma is a very hot gas composed of electrons and ions. A variety of interactions may occur in a plasma which can lead to the production of X-rays. A free electron which passes close to an ion will have its path curved by the attraction of the positively charged nucleus, and, since curving the path of the electron is equivalent to decelerating it, the electron radiates some energy in the form of a photon. The wavelength of the photon produced depends partly on the energy of the electron; an electron with an energy of a few keV may emit an X-ray. This process is known as Bremsstrahlung (the German word for 'braking radiation') or, since the electron starts free of the ion and retains its freedom afterwards, as free–free radiation. An electron which encounters an ion as it travels through a plasma does not always escape; sometimes it is captured into one of the ion's outer electron shells. The energy of the electron in the shell is less than the energy it had when it was moving through the plasma, and the difference in energy is released as a photon. This is known as free–bound radiation. The free electrons in the plasma all have different energies, so each photon emitted as free–free or free–bound radiation will have a slightly different energy, and the result will be a continuous spectrum, that is, a mixture of photons with a wide range of different energies.

There is also a process which leads to the emission of photons with a specific energy,

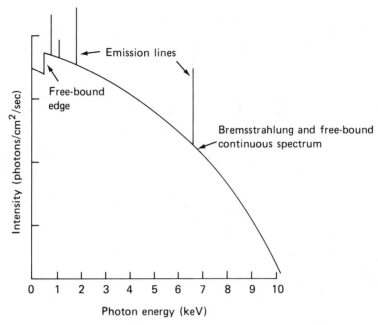

Fig. 2.1 The X-ray emission from a high temperature gas including Bremsstrahlung, free-bound and line radiation. (Reproduced, with permission, from J. L. Culhane and P. W. Sanford, *X-ray astronomy*, Faber and Faber 1981).

that is, a 'spectral line'. Line emission occurs when a free electron or ion interacts with an electron already bound to an ion and 'excites' the bound electron, pushing it up to a higher energy level within the ion. The excited electron is unstable in its new position and soon falls back to its original level, or 'ground state'. As it does so it emits a photon, the energy of which is equal to the difference in the energy between the two electron states. Since these energies are fixed, every photon emitted by a certain transition has the same energy, and the result is a spectral line. A consequence of this process is that each chemical element, and each of its ions, has a unique series of spectral lines corresponding to its different energy levels. These lines can be used to identify the elements present in astronomical objects. In a plasma where the material is already highly ionised, these energy differences correspond to X-ray photons.

Thus a hypothetical X-ray spectrum from a plasma might look like Fig. 2.1, in which the Bremsstrahlung and free–bound emission produce a smooth curve on which a few emission lines are superimposed. Sharp edges such as the one at about 0.5 keV, are called free–bound edges and arise at energies equal to the energies of atomic shells into which incoming electrons are being captured. The shape of the underlying curve is dependent on the temperature of the plasma, since this determines the number and energy distribution of the free electrons and the ionisation states of the atoms, but it is not the same shape as that produced by black body radiation.

2.2 X-RAYS FROM THE SUN

X-rays from the Sun were first detected with rocket experiments, the simplest of which were pieces of film shielded from visible light and ultraviolet by beryllium or aluminium foil. By varying the thickness of the foil it was possible to detect X-rays of different energies; X-rays of all wavelengths would penetrate a thin foil, but as the thickness of the foil was increased, only the more energetic photons could reach the film. As well as the pioneering launches from New Mexico, this type of detector was also carried on a number of British Skylark sounding rockets during 1959 and the early 1960s.

A more sophisticated technique for X-ray observations of the Sun was developed by Herbert Friedman's group from the US Naval Research Laboratory (NRL) in Washington. The NRL group used rocket borne Geiger counters, and, since similar counters form the basis of numerous X-ray detectors, it is worth examining how such a counter works. The counter consists of a sealed chamber containing an inert gas such as argon, neon, or xenon, plus some methane. One or more positively charged (anode) wires run across the chamber, and each wire is maintained at a potential of a few thousand volts. If an X-ray enters the chamber and strikes a gas atom the X-ray may be photoelectrically absorbed and eject an electron from the atom. The electron is attracted towards the anode wires running across the chamber and travels through the gas, undergoing further collisions and producing additional electrons as it does so. In about one millionth of a second a cloud of free electrons is formed which falls onto the anode wire and produces a current pulse which can be amplified and detected. If the potential difference across the electrodes is large, the current pulse produced is independent of the energy of the incoming photon (provided of course that the photon has sufficient energy to cause the ionisation in the first place), and the detector is said to saturate.

A Geiger counter provides no information about the energy of the photons which it detects, so the NRL group fitted their counters with thin foil 'windows' to restrict the range of wavelengths which could trigger the detector. By using a number of counters, each sensitive to a different energy band, it was possible to get an idea of how the Sun's X-ray emission varied with wavelength, that is, to determine the Sun's X-ray spectrum.

There is, however, a better way of getting information about the energies of X-ray photons than using a number of Geiger counters with foil windows of varying thicknesses. If the potential across the anode wires in the counter is reduced to a suitable level, then the strength of the current pulse produced is proportional to the energy of the photon which triggered the original electron avalanche. A counter operated in this way is known, logically enough, as a proportional counter. Proportional counters still need foil windows to contain the gas, but the foil is chosen to admit photons with a wide range of energies, and electronics are used to determine the spectrum of the X-rays detected.

Unfortunately, proportional counters can be triggered by charged particles as well as by X-ray photons. This means that a proportional counter used for astronomy is constantly recording a background of detections due to cosmic rays and particles trapped by the Earth's magnetic field. To minimise this, X-ray satellites must avoid the Earth's radiation belts as much as possible, and data analysis techniques must take into account the background noise. Often, separate counters are used to detect charged particles, and only signals which do not occur at the same time as particle hits are regarded as genuine X-ray detections. These are known as anticoincidence shields since they cause the experiment to ignore signals which occur 'coincidentally' in the two detectors.

Friedman's group, and a few others, continued to develop X-ray counters and to make X-ray observations of the Sun. They confirmed that the solar corona was composed of a million degree plasma and that X-rays were produced in active regions on the Sun's surface. Some of the techniques which they used were ingenious and included multiple rocket launches from the deck of a US Navy destroyer (USS *Point Defiance* on 12 October 1958) and launching rockets from balloons (known as Rockoons) to carry their instruments to greater altitudes. Unfortunately, X-ray astronomy was almost a victim of their success; the NRL results showed that X-rays from the Sun were only detectable because the Sun was so close and the chances of detecting another star, millions of times further away, seemed remote. The NRL group was unlucky; extrasolar X-ray sources did exist, but they were just too faint to be detected by their instruments; the honour of discovering an X-ray star was to fall to another group.

2.3 THE FIRST X-RAY 'STARS'

The first X-ray source beyond the solar system was discovered in June 1962 by a group of scientists working for American Science and Engineering (AS&E) Inc. The AS&E group, led by Riccardo Giacconi, were carrying out research work for the US Air Force Cambridge Research Laboratory. They had previously launched two X-ray astronomy rockets, both of which had failed for technical reasons, and had undertaken classified work relating to the detection of atmospheric nuclear explosions. The stated purpose of the June flight was to determine if the Moon might be a source of X-rays produced when the solar wind struck the lunar surface, but it is clear that behind this official objective was the hope that the rocket might discover cosmic X-ray sources as well.

The AS&E group launched an Aerobee rocket equipped with three proportional counters. Each counter was only a few centimeters in size and scanned across the sky as the rocket rotated around its long axis. The flight was a success, discovering a bright X-ray source in the direction of the constellation Scorpius. In fact the source was so bright that the first reaction of the watching scientists was that there was a problem with the instruments; the source was much stronger than even the optimists had expected. After two months of data analysis it was clear that the source was real, and this was established beyond doubt by a further flight in June 1963. In the meantime the AS&E group found hints of two other X-ray sources in data from a rocket launched in October 1962 (the Scorpius source was unobservable from the White Sands launching site in October and could not be verified on this flight), and in April 1963 Friedman and co-workers found a second definite X-ray source, this time in the direction of the Crab Nebula.

Over the next few years several other groups entered the X-ray astronomy field, including the Lockheed research laboratory, the Massachusetts Institute of Technology (MIT), and Leicester University. All were using rocket launched detectors, but methods of stabilising the rockets, and the technology of the detectors, were being steadily improved. By 1970 the number of X-ray sources known had risen to more than forty, some of which were now being identified with objects already known to other branches of astronomy such as supernova remnants (the remains of exploded stars), the centre of our own Galaxy, and one or two objects apparently connected with external galaxies. There also seemed to be a uniform diffuse X-ray emission of unknown origin detectable over the entire sky. The precise nature of most of these X-ray sources remained a mystery, and, because of the low angular resolution of X-ray detectors, many could not even be identified with known objects. X-ray astronomers everywhere were awaiting the launch of the first satellite able to survey the entire sky in X-rays.

2.4 X-RAY SURVEYS

2.4.1 Initial attempts

The limited observing time available during a rocket flight (5–10 min) made progress in X-ray astronomy slow, and it was clear that orbiting instruments were needed if the science was to advance. X-ray detectors were carried on the NASA Orbiting Solar Observatories (OSO); satellites which consisted of a Sun pointing sail section and a wheel which rotated beneath the sail to provide gyroscopic stability. Some instruments were carried in the sail, and so viewed the Sun continuously, whilst others were fitted to

The OSO-3 satellite under test. OSO-3 carried a small X-ray experiment in the spinning wheel section, here seen below the hemispherical sail which carried the solar cells. The small spheres contain gas for the attitude control system and swung outwards when the satellite was in orbit. (NASA).

the wheel and scanned across the sky (and the Sun) as the wheel rotated. OSO-1 and 2 were launched in 1962 and 1965 respectively, and operated for several months.

The OSOs were, however, designed for solar, not deep space, observations, and luck did not favour the first attempts to use satellites for extra-solar X-ray astronomy. The Lockheed research laboratory placed a proportional counter designed to study stellar X-ray sources on the first Orbiting Astronomical Observatory, but this spacecraft failed before any observations could be made (see section 4.4.1). Extra-solar X-ray instruments were also carried on the OSO-C satellite, but the third stage of the Delta launcher ignited before it had fully separated from the second stage, and OSO-C crashed into the Atlantic Ocean. In an ironic jest, an aide to the project manager referred to the crashed satellite as 'OSO-Sea'. The replacement satellite, named OSO 3 once in orbit, made some limited extra-solar X-ray observations with a small counter (area about 10 sq cm) mounted in the satellite's wheel section.

The European Space Research Organisation's satellite Iris (ESRO-2B), which was designed mainly for studies of the X-ray and particle emissions from the Sun, is credited with extra-solar X-ray observations as are the US Air Force's Vela satellites. The role of the Vela series, described in section 3.4.1, was to detect nuclear detonations from space, but several in the series carried X-ray astronomy instruments. The most notable was Vela 5B (also known as Vela 10) whose X-ray detectors operated from May 1969 to June 1979. During its very long lifetime Vela 5B was able to monitor the entire sky in the energy range 3–12 keV and to build up an important record of a variety of transient X-ray sources. Some of the Vela observations are particularly important because they cover periods when no other X-ray satellites were operating. Vela 5A (Vela 9) and Vela 6A and 6B (Vela 11 and 12) also detected extra-solar X-ray sources. All of these spacecraft, and some Soviet Cosmos satellites carrying X-ray detectors, have now either re-entered or have been deactivated. They are listed in Table 2.1 for completeness. For a review of the Vela 5B mission see the bibliography.

The Vela 5 satellite in orbit (simulated image) (Los Alamos National Laboratory).

Table 2.1.
Satellites with extra-solar X-ray experiments

Name	Launch date	Launch vehicle	Orbit† Perigee (km)	Apogee (km)	Inclination°	Notes
OSO-C	25 Aug 1965	Delta	—	—	—	Launch failure
OSO-3	8 Mar 1967	Delta	339	345	39.2	Mostly solar experiments
Cosmos 208	21 Mar 1968	A1/2	129	190	65.0	Military
Cosmos 215	19 Apr 1968	B-1	162	265	48.5	8 visible, UV and X-ray experiments
ESRO-2B	17 May 1968	Scout	205	667	97.2	European mission
Cosmos 262	26 Dec 1968	B-1	163	508	48.5	Military
Vela 5A	23 May 1969	Titan 3C	94 052	128 529	56.4	Vela 9 } Military dual launch
Vela 5B	23 May 1969	Titan 3C	89 999	133 011	56.2	Vela 10 } 180° apart
OSO-6	9 Aug 1969	LTTA-Delta‡	324	328	33.0	Mostly solar experiments
Vela 6A	8 Apr 1970	Titan 3C	106 367	116 056	54.9	Vela 11 } Military dual launch
Vela 6B	8 Apr 1970	Titan 3C	105 560	117 188	54.8	Vela 12 } 180° apart
SAS-1	12 Dec 1970	Scout	324	350	3.0	Uhuru, Explorer 42. X-ray sky survey
Cosmos 428	24 Jun 1971	A-2	128	160	51.7	Military
OSO-7	29 Sep 1971	LTTA-Delta‡	201	355	33.1	Mostly solar experiments
TD-1A	12 Mar 1972	Delta	531	539	97.5	X-ray experiment failed
OAO-3	21 Aug 1972	Atlas Centaur	729	739	35.0	UV observatory, secondary X-ray experiment
Skylab-3	28 Jul 1973	Saturn 1B	157	237	50.0	S150 experiment in S-IVB rocket.
ANS	30 Aug 1974	Scout	256	1098	98.0	Ist Netherlands satellite
Ariel 5	15 Oct 1974	Scout	513	557	2.9	UK/US X-ray observatory
Salyut 4	26 Dec 1974	D-1	337	350	51.6	Space station. X-ray experiments 'Filin' and RT-4 (0.16–0.3 keV) telescope.

Name	Date	Launcher				Notes
Arabhata	19 Apr 1975	C-1	530	536	83.0	India/USSR. Failure.
SAS-3	7 May 1975	Scout	506	513	3.0	Explorer 53
OSO-8	21 Jun 1975	Delta	467	475	32.9	Mostly solar experiments
HEAO-1	12 Aug 1977	Atlas Centaur	424	444	22.7	All sky survey
HEAO-2	13 Nov 1978	Atlas Centaur	355	364	23.5	Einstein observatory
Corsa-B	21 Feb 1979	Mu-3C	473	494	29.9	Japanese soft X-ray Mission (Hakucho)
Ariel 6	2 Jun 1979	Scout	562	600	55.0	Cosmic-ray/X-ray
Astro-B	20 Feb 1983	Mu-3S	488	503	31.5	Tenma
EXOSAT	26 May 1983	Delta	356	191 581	72.5	ESA observatory
Spartan 1	17 Jun 1985	Space Shuttle (Discovery STS-51G)	356	352	28.4	Free flying experiment platform (see section 8.2)
SL-2 XRT	29 Jul 1985	Space Shuttle (Challenger, STS-51F)	311	319	49.5	Spacelab-2 Coded-mask Telescope
Astron 1	23 Mar 1983	D1e	2000	200 000	51.5	SKR-02M experiment
Astro-C	5 Feb 1987	Mu-3S	510	673	31.1	Ginga
Kvant	31 Mar 1987	D-1	344	363	51.6	Science module docked to Mir space station

† Since satellite orbits change owing to atmospheric drag etc., orbital parameters quoted by different sources may vary.
‡ Long Tank Thrust Augmented

2.4.2 SAS-1 (Uhuru)

The first satellite dedicated to X-ray astronomy was the first in a series of Small Astronomy Satellites (SAS), and was also known as Explorer 42. SAS-1 was conceived by Giacconi and his group at AS&E as a follow-on to their rocket experiments. The satellite was originally called the X-ray Explorer and was proposed to NASA in April 1964 and approved in late 1966. Development of the scientific instruments was left in the hands of AS&E, but the spacecraft was designed and built by the Applied Physics Laboratory of Johns Hopkins University. The project was directed by the Goddard Spaceflight Center in Maryland.

The spacecraft consisted of two main elements. One element was a standardised cylindrical body about 0.6 m in diameter and 0.6 m high containing the control and data handling systems. Four solar panels, each 4 m long, were attached near the bottom of the body and were deployed after the satellite reached orbit. The satellite rotated about its long axis every 12 minutes, allowing the X-ray detectors to scan across the sky. Stability was provided by a momentum wheel spinning at 2000 rpm. A magnetorquer

The SAS-1 (Uhuru) satellite is prepared for launch. (NASA).

was used to reduce the spin rate for special observations and to precess the spin axis to re-orient the satellite's viewing direction.

The second element was an instrument package mounted on top of the cylindrical body. This contained two argon filled proportional counters both sensitive to X-rays in the range 2–20 keV and based on detectors developed by AS&E for rocket flights. To improve their angular resolution, and hence get a more accurate position of sources detected, the counters were fitted with collimators, mechanical devices made of thin metal plates that restrict the field of view of the detector. (A collimator acts rather like a blinker over a horse's eye.) One detector was fitted with a collimator with a field of view of 5 × 5 degrees, the other with one covering 5 × 0.5 degrees. The detectors observed at right angles to the satellite spin axis, and so both scanned the same strip of sky in turn as the satellite rotated. Sun and star sensors mounted above the counters were used to assist in determining the pointing direction of the satellite and thus to locate the position of X-ray sources. Data from the instruments were stored on an onboard tape recorder and played back at 30 times normal speed when the satellite passed over a NASA ground station at Quito, Equador.

To take advantage of the boost provided by the Earth's rotation, and to place the satellite in an orbit over the Equator which would minimise interference from the radiation belts, SAS-1 was launched from the Italian San Marco launch platform. The platform stands off the coast of Kenya (longitude 40° 12′ 47″ East, latitude 2° 56′ 40″ South) and is supported on 20 steel legs. It is equipped to launch NASA's four stage, solid fuelled Scout rocket and includes an air conditioned building for integration and checkout of the launcher and its payload. Control is provided from the Santa Rita platform, a converted oil rig moored about 500 metres away.

The satellite was launched on 12 December 1970. By coincidence this was the 7th anniversary of Kenya's independence, and the satellite was unofficially named Uhuru, the Swahili word for freedom. The name was not recognised by NASA at the time, but has since come into general use. The original observing plan was to scan a single strip of sky for 24 hours then activate the magnetorquer to change the spin axis and observe a different region. In this way the entire sky could be covered in about 2 of the 6 months the satellite was expected to operate. As the mission progressed this observing plan was modified to observe specific sources in more detail. A comprehensive review of the flight of Uhuru is given in *The X-ray universe* by W. Tucker & R. Giacconi and further details are to be found in *Frozen star* by G. Greenstein (see *Bibliography*), so only brief highlights of the scientific results will be given here.

Probably the greatest success was the discovery of the X-ray binaries. An example of these was the object Cen X-3, the third X-ray source discovered in the constellation of Centaurus. Rocket observations, made before Uhuru, had discovered this source and had hinted that it was variable. Uhuru found that not only was there a regular five second variation in the X-ray output from Cen X-3, but that the period between these variations was changing in a regular way. The puzzle was resolved when it was realised that the X-ray source was orbiting another star, and as the X-ray source approached and receded from the Earth, the Doppler effect systematically changed the frequency of the X-ray pulses. This idea was supported by the fact that every two days or so the source disappeared for a few hours as the X-ray star vanished behind its companion. The companion star was eventually identified by optical astronomers and proved to have the same binary period as the X-ray source, confirming the connection.

Studies of Cen X-3, and a similar object Hercules (Her) X-1, eventually revealed that the X-ray source was a rotating neutron star which was accreting material from its companion. The material blown off the companion star is funnelled down onto the neutron star's surface by powerful magnetic fields. The X-rays are produced either in the funnel of material, or as the material crashes onto the surface of the neutron star. The regular pulsations are caused by the rotation of the neutron star flashing this 'hot spot' in and out of sight, rather like the flashes of a lighthouse beam.

Another early target of the mission was the source known as Cygnus (Cyg) X-1. Different groups had observed Cyg X-1 from rockets and had obtained different results, suggesting that the source might be variable. Variability from week to week was soon confirmed by the Uhuru data which also suggested that as well as this long term variability, Cyg X-1 varied on very short timescales, possibly as short as a few milliseconds. These results have now been confirmed, and Cyg X-1 is now believed to be a black hole, orbiting the giant blue star HDE226868 in a bizarre binary system.

As well as these exciting discoveries, an important objective of Uhuru was to make a survey of X-ray sources, and, about two years after launch, a catalogue of 150 objects was published. After additional data analysis, another list, known as the 4U (i.e. the 4th Uhuru) catalogue, appeared in 1978. This contained 339 sources, many of which were supernova remnants, X-ray binaries, and globular clusters. The catalogue also listed some sources outside our Galaxy such as X-ray stars in the Magellanic Clouds, the Andromeda galaxy, and a few other spiral galaxies. The catalogue also contained a number of active galaxies such as Seyfert galaxies and quasars. Uhuru also discovered that several galaxy clusters and superclusters have extended regions of X-ray emission associated with them. This diffuse emission is attributed to the presence of very tenuous, 100 million degree, gas spread between the galaxies in the cluster. The production of these catalogues, and the discoveries which preceded them, marked Uhuru as a triumphant achievement.

This success was not achieved without its share of problems; after only two weeks in orbit the satellite was in trouble. Engineers deduced that Uhuru had overheated, and

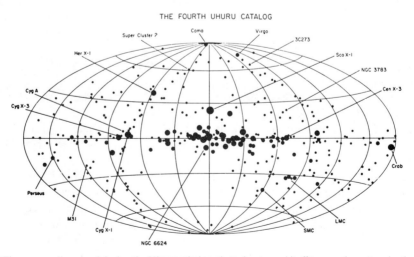

The sources discovered during the Uhuru mission plotted onto an Aitoff (or equal area) projection of the sky. (Harvard–Smithsonian Center for Astrophysics)

slewed it so that only a small area faced the Sun, and within hours the satellite had cooled and was back in action. A month later the tape recorder jammed, meaning that data could be received only during the 8 minutes per orbit that Uhuru was visible from Quito. The rest of the data were irretrievably lost. NASA brought other ground stations into action, each recording 8 minutes' data per orbit, and tapes were sent to the Goddard Spaceflight Center and spliced together before being forwarded to AS&E. This drastic measure meant that about half the data taken by the satellite could be saved for analysis. Next, the transmitter began to fail, the signals became weaker and weaker, and contact was almost lost. This problem was solved by accident; NASA engineers experimenting with the satellite for a quite different purpose commanded it to swing to point at the magnetic North and somehow switched the transmitter into another operating mode. For no obvious reason the transmitter sprang back to life and the mission continued. Uhuru finally ceased operating on 12 January 1974 and re-entered on 5 April 1979.

2.4.3 Other early X-ray experiments

While Uhuru was still operating, several other satellites carrying X-ray detectors were launched.

Cosmos 428, launched 24 June 1971, was a military satellite which also carried X-ray astronomy experiments. These included a scintillation spectrometer sensitive to X-rays with an energy greater than 30 keV and restricted by collimators to a field of view of

The OSO-7 spacecraft. The solar panels on the sail remain pointed at the Sun while the wheel section below rotates, scanning the experiments across the sky. (NASA).

$2 \times 17°$ and what was described as an X-ray telescope operating in the 2–30 keV range. During a brief mission Cosmos 428 detected a number of X-ray sources, many of which were identified with sources in the Uhuru catalogues.

Orbiting Solar Observatory 7 carried two X-ray experiments, one to survey the sky in the 1–60 keV range with an angular resolution of about 1 degree, the other designed to measure the position, intensity, and spectrum of X-ray sources in the 10–550 keV range. During launch a malfunction of the second stage of the Delta rocket sent OSO-7 into the wrong orbit and left the satellite spinning with its wheel section facing the Sun and its solar panels in shadow. With no power available from the solar panels, the satellite's battery immediately began to run down as NASA engineers raced to correct the problem. With battery voltage at 17 V, 0.2 V above the probable failure threshold, the satellite was re-oriented and power began to flow from the solar panels. After these initial problems OSO-7 operated for about 20 months and produced a catalogue of observations which included studies of point sources and of the diffuse X-ray background.

Next to carry X-ray instruments into orbit was the European TD-1A satellite described in section 4.5. Although basically an ultraviolet observatory, TD-1A was also equipped with a proportional counter (collecting an area of 100 sq cm) designed to measure the spectra of cosmic X-ray sources in the 3–30 keV range. Unfortunately, when the experiment was activated it caused abnormal readouts in the satellite's telemetry channels, and the experiment was switched off as a precautionary measure.

The run of problems was broken by the last Orbiting Astronomical Observatory, OAO-3 (section 4.4.4). Named Copernicus, OAO-3 was an ultraviolet observatory, but it also carried an X-ray instrument, provided by British astronomers, as a secondary payload. The original Anglo–US agreement called for 10% of the observing time to be used for X-ray observations, with the remainder allocated to ultraviolet studies.

The OAO-3 instrument consisted of four X-ray telescopes. Three of them used parabolic mirrors to direct incoming X-rays onto very small detectors, less vulnerable to background radiation than counters with a large collecting area. The fourth, for the hardest X-rays, was a collimated proportional counter. The instrument was proposed in 1963, but had to wait until 1972 before OAO-3 was launched. For many experiments this nine year delay would have been disastrous, but for the British group it proved a happy circumstance. By the time Copernicus had been launched, Uhuru had already identified a number of promising targets for just the kind of detailed observations for which the OAO series was designed. In fact the Copernicus X-ray instrument proved so successful that the American project scientist Lyman Spitzer agreed to double the time allocated to X-ray observations, even though this meant reducing the number of observations with the ultraviolet telescope.

The accurate pointing capability of Copernicus, together with its long operating lifetime (the satellite was not deactivated until 1981), meant that the X-ray package was able to make many important discoveries. For example, Copernicus made long observations of Cygnus X-3 and demonstrated conclusively that the object varied with a 4.8 hour period. Similar results had been found by Uhuru scientists, but the limitations imposed by the continuous rotation of Uhuru made their data analysis much more difficult. Astronomers using Copernicus also made detailed observations of the black hole candidate Cygnus X-1, studied the Crab Nebula, and used the fine pointing capabilities of the OAO to make detailed maps of supernova remnants in the constellations of Cassiopeia and Puppis. Copernicus also discovered the existence of a new class of pulsating X-ray sources with periods of a few minutes. These so called 'slow rotators' are believed to be neutron stars in binary systems. Observations were also made of extragalactic sources, and it was discovered that the unusual galaxy Centaurus A (also known as NGC 5128) was four times brighter in 1975 than when it was first observed by Uhuru. Over the next few months Centaurus A faded back to its original X-ray brightness. This was the first example of a variable X-ray source beyond our galaxy.

A further success came with an experiment flown as part of the Skylab space station project. Unlike the ultraviolet telescopes described in section 4.6.2, the S150 X-ray experiment was not carried in the space station but was fitted to the 3rd stage of the Saturn 1B rocket that launched the second crew towards the orbiting Skylab in July 1973. S150 consisted of three proportional counters, total collecting area 1500 square centimetres, attached to the inside wall of the instrument unit which was itself mounted atop the SIV-B upper stage of the Saturn rocket. The instrument was sensitive to photons with wavelengths between 4–10 nm, a region which marks the crossover from very soft X-rays to the extreme ultraviolet (see section 4.13). To permit the entry of such low energy X-rays the front window of the proportional counter consisted of a plastic sheet only 2 μm thick. As the counter gas, a mixture of argon and methane, escaped through the window it was maintained at the correct density by a sophisticated pressure regulator linked to an onboard gas supply. Collimators on each counter defined three intersecting regions of the sky (each approximately $2 \times 20°$) as an aid to locating any point sources that might be observed.

After the astronauts had separated their Apollo capsule from the SIV-B stage the S150 experiment was deployed from its protective housing and activated. The entire SIV-B stage underwent a series of preprogrammed manoeuvres, scanning at about 1° per 15 seconds, to allow the instrument to sweep across selected regions of the sky. The pointing direction was determined during data processing, using the inertial guidance system of the SIV-B stage combined with information from two visible star sensors which formed part of the experiment. Data were stored on a tape recorder and replayed to suitable ground stations when possible.

The object of the experiment was to determine if the soft X-ray background previously detected from sounding rocket flights was due to large numbers of discrete X-ray sources too faint to be resolved individually with smaller detectors or was truly diffuse. The experiment operated successfully (until exposure to sunlight melted the thin plastic window), and it was found that even though many stars passed through the field of view of the instrument no point sources of 4–10 nm radiation could be identified. This result showed that the soft X-ray background was indeed diffuse, and probably arose in hot gas spread throughout the interstellar medium.

In 1974 another nation entered the X-ray astronomy field with the launch, by an American Scout rocket, of the Dutch ANS (Astronomische Nederlandse Satelliet) spacecraft. The ANS was three axis stabilised and was built by an industrial consortium of Fokker-VFW and Philips. Ultraviolet and soft X-ray instruments were provided by the Universities of Groningen and Utrecht respectively, and an MIT/AS&E group supplied a hard X-ray instrument. The ANS used reaction wheels unloaded by magnetorquers for attitude control, and could be pointed at a target and then stabilised with an accuracy of about 20 arc seconds. The spacecraft used an onboard computer to

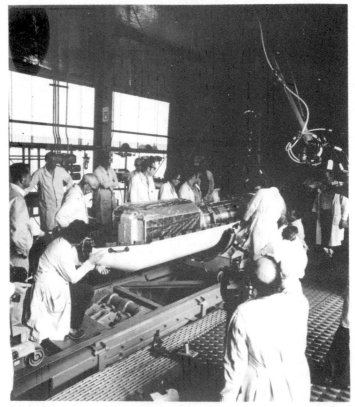

The Netherlands ANS satellite is integrated with its Scout launch vehicle in 1974. (NASA).

store commands for twelve hour periods and operate automatically between ground contacts with the European Space Agency's Operations Centre (ESOC) at Darmstadt, W. Germany.

The ANS was launched from the United States Western Test Range in California on 30 August 1974. Unfortunately because of a failure of the launcher's guidance system the satellite did not reach its planned 500 km, sun-synchronous, polar orbit (inclination 97.5°), but was instead placed in a highly elliptical orbit ranging from 280 to 1150 km. This orbit caused various problems with background radiation and complicated the planning of the observing programme, but these difficulties were gradually overcome as the mission progressed. ANS operated until it re-entered in 1977; it made two important discoveries about X-ray sources within our galaxy.

The first of these was that cool dwarf stars could emit a detectable amount of X-rays. As noted earlier, rocket experiments had shown that the Sun was a comparatively weak X-ray source, and the detection of X-rays from solar type stars was not expected without instruments much more sensitive than those on ANS. Despite this, ANS detected X-rays from a class of dwarf stars called flare stars (the nearest flare star is also the closest star to the Sun, Proxima Centauri). These stars, also called UV Ceti stars, undergo brief, but sometimes spectacular, changes in their optical brightness, and, not surprisingly, it was found that the X-ray emission seemed to be related to these optical bursts. Since the surfaces of these stars are cool, the X-rays undoubtedly originate in the stellar coronae. The flares are probably related to the lower mass, and hence different internal structure, of these stars compared with the Sun. Low mass stars have a structure in which convection, rather than radiation, brings energy from the central regions to the surface, and magnetic disturbances which start deep below the surface are magnified by turbulence as they travel upwards through the star. When the disturbances reach the surface they cause the formation of huge starspots. Magnetic energy stored in these regions is suddenly released and converted into energetic particles and radiation in a process analogous to, but many times more powerful than, a solar flare.

The second discovery was made by Johnathon Grindly of the Harvard Smithsonian Center for Astrophysics. He found an X-ray star that was giving out sudden, very brief, bursts of X-rays in the 1–10 keV range. So dramatic were these bursts that this star gave out more energy in X-rays during a few seconds than our Sun radiates at all wavelengths over a period of weeks. Once alerted to this unexpected phenomenon, astronomers using data from other satellites, including some launched after ANS such as Ariel 5, OSO-8, and SAS-3, soon discovered other 'X-ray bursters'. Subsequent observations showed that in some of these sources the X-ray bursts occurred at almost the same time as surges in the optical brightness of the star. The bursters are now thought to be binary systems containing a neutron star and a normal companion. Material from the companion star, guided by magnetic fields, spirals onto the neutron star and accumulates on the surface. When a critical concentration is reached this hydrogen rich material undergoes a massive thermonuclear explosion called a helium flash which produces a sudden burst of energy. X-ray bursters are thus similar to the X-ray binaries discovered by Uhuru; the difference is that the normal X-ray binaries give off energy in a steady stream as material falls onto the neutron star, but the bursters produce energy in a much more dramatic fashion with periods of relative calm in between.

2.4.4 Ariel 5

The next satellite devoted entirely to X-ray astronomy was Ariel 5. The Ariel series began in the early 1960s as a joint US/UK programme, and the first four satellites were devoted to ionospheric physics. As the series continued, British astronomers urged that a satellite be devoted to X-ray astronomy, and contracts for the development of the Ariel 5 spacecraft were placed in 1970.

Ariel 5 was a spin stabilised, drum shaped, satellite 86.5 cm long and 95 cm in diameter. Solar cells covered most of the curved surface of the drum, except for apertures for two scientific instruments. A further four experiments viewed out of the top of the drum, parallel to the spin axis. The side mounted instruments scanned the sky as the satellite rotated; the others were pointed by using propane gas thrusters to precess the satellite until the spin axis was aimed in the correct direction. Ariel 5 also carried a magnetorquer to make small adjustments to the direction of the spin axis. The maximum rate of change with the magnetorquer was only about 0.1 degree per orbit, but it proved vital to the mission because it allowed observations to continue long after the propane gas was exhausted.

The two side mounted instruments were a Sky Survey Instrument which used large

The Ariel 5 satellite. (Marconi Space Systems and SERC)

proportional counters sensitive to X-rays in the range 1.5–20 keV, and an American supplied all sky monitor which had a large field of view and was able to detect transient events and alert ground controllers so that follow up observations could be made. The all sky monitor, and the launch, were the United States' contribution to the bilateral mission. Some of the pointed instruments used a number of interesting observational techniques described below.

The Rotation Modulation Collimator was used to measure the positions of X-ray sources with much greater accuracy than is possible with an ordinary proportional counter. An RMC uses parallel grids in front of a conventional X-ray detector; if the collimator is rotated, or if the instrument is looking along the axis of a spinning satellite, the rate at which X-rays are detected varies according to the angle between the source and the collimator. The modulation of the X-rays can be detected during data analysis and the position of the source calculated with considerable precision (less than 1 arc minute for Ariel 5). The Ariel 5 instrument also used a collimated optical photomultiplier to provide additional information on where the satellite was pointing during X-ray measurements.

The second pointed experiment was a Bragg Crystal Spectrometer used to search for spectral lines in bright X-ray sources. These spectrometers work by using a crystal, in which the planes of atoms in the crystal lattice are separated by an distance d, to reflect photons arriving at a very shallow angle. Reflection occurs only if the angle of incidence (the Bragg angle θ) and the wavelength of the photon (λ) are related by the expression $2d . \sin \theta = n\lambda$, where n is an integer. Although capable of high spectral resolution, these instruments are not very efficient and can only be used to study bright sources. The Ariel 5 instrument was sensitive to photons between about 2 and 9 keV.

High energy photons, stretching almost into the gamma ray region, cannot be studied with gas filled proportional counters because the probability of a photon being absorbed by a gas atom decreases as its energy increases. To observe these high energy

X-ray Ariel 5 carried a scintillation counter. This type of instrument uses a crystal of caesium iodide to detect the X-ray photons, which cause a brief flash of light as they pass through the crystal. The light flashes are detected by a sensitive photomultiplier, and the intensity of each flash is proportional to the energy of the incoming photon. The Ariel 5 scintillation counter was sensitive to energies between about 26 and 1200 keV and could locate sources with an accuracy of about 2°.

The fourth experiment aligned with the spin axis used a multiwire proportional counter to measure the X-ray spectra of individual sources in the 1.4–30 keV range.

Ariel 5 was launched from the San Marco platform on 15 October 1974 and placed in a near circular orbit inclined at 2.9° to the Equator. After engineering checks, X-ray observations commenced on 22 October. Operations were controlled from the Appleton Laboratory near Slough, UK, which was in touch with scientists throughout the UK and, via a complex communications link (from the UK to Spain, across the Atlantic via satellite to the USA, and hence to tracking stations at Quito, Ecuador, and Ascension Island) with the satellite itself.

Ariel 5 made many contributions to X-ray astronomy, and only a few of them will be mentioned. The Sky Survey Instrument produced several important catalogues of locations, spectra, and identifications of X-ray sources. The 2A catalogue, published in 1978, covered regions of the sky more than 10° from the galactic plane and included 150 sources, 40 of which were new discoveries. The final, 3A, catalogue covered the whole sky and contained 250 sources.

Ariel 5 observed a number of transient sources and was able to determine their positions accurately enough for some to be identified. One object in the constellation of Monoceros brightened until, for a while, it was the brightest X-ray source in the sky. The positions provided by Ariel 5 and other satellites allowed the source to be linked with a nova occurring about 6000 light years from the Sun. Ordinary novae occur in binary systems containing a compact object like a white dwarf and a normal star, but Nova Monoceros 1975 was so far away that a normal nova could probably not produce enough energy to account for the X-ray flux detected. Since the power produced by accretion onto a compact object increases as the ratio of mass-to-radius of the accreting star, it was suggested that the compact object in Nova Monoceros 1975 was either a neutron star or a black hole. Evidence that Nova Monoceros involved a black hole came when Cygnus X-1, another black hole candidate, underwent a similar, although less spectacular, outburst. Other transients were found to vary regularly with periods of a few minutes. One of these reached its peak brightness on 25 December 1974 and was known for a while as Cen X-mas until the rather more prosaic catalogue identification of A1118–61 was applied to it.

Ariel 5 also observed the X-ray bursters discovered in ANS data and established that at least some of them emit a steady background of X-ray between bursts. Observations of a very powerful burster called MXB1930–335 (the name identifies it as an MIT survey X-ray Burster at celestial coordinates Right Ascension 19 h 30 m Declination −33.5°) showed that the bursts, each a few seconds long, came in patterns which seemed to have a period of between 10 to 20 seconds. Very hard X-rays were also detected from the source, suggesting a link with the gamma ray bursters discussed in section 3.4.

Ariel 5 observations discovered the presence of an X-ray spectral line at 6.7 keV in the supernova remnants Cassiopea A and Tycho. This is an emission line of an iron atom with all but two of its 26 electrons removed (the iron is said to have an oxidation state of 25, so the line is usually referred to as FeXXV), and its detection was important for two reasons. Firstly it confirmed that the X-ray emission from these old supernova remnants was due to the presence of very hot plasma, caused as the shockwave from the exploding star crashes into gas in the interstellar medium and heats it up, and was not due to synchrotron radiation caused by electrons spiralling around in magnetic fields as in the case of the Crab Nebula. Secondly it allowed the temperature of the shock heated gas and the abundance of iron in the gas to be estimated. The iron abundance is of particular interest to astronomers studying stellar evolution since iron is produced deep inside stars and cannot normally be detected.

Beyond our Galaxy, Ariel 5 found spectral lines of Fe XXV and Fe XXVI in the emission from clusters of galaxies, important because the iron lines confirmed that the X-rays originated in a 10–100 million degree gas throughout the cluster. Ariel V data also showed that the X-ray power of a cluster increases with the number of galaxies, and the clusters which emit the most X-rays are relatively devoid of spiral galaxies. This suggests that the gas is stripped from the spiral galaxies as they pass through the denser central regions of the galaxy cluster, and, with their star forming regions disrupted, the spirals evolve into elliptical galaxies.

Ariel 5 also identified a number of extragalactic sources with Seyfert galaxies and confirmed that in some of these active galaxies the X-ray emission could vary over periods as short as one day. This means that the X-ray emitting regions in these galaxies must be relatively small, and that in turn means that they must have very powerful energy sources. A favoured suggestion is that supermassive black holes lurk in the nuclei of Seyfert galaxies and generate X-rays by tearing apart and then devouring stars which wander too close to the nucleus. Surprisingly, the rate at which stars must be consumed in this way turns out to be quite low, only about one or two stars per year.

Ariel 5 was still operating when it re-entered the atmosphere on 14 March 1980. It had been a tremendous success.

2.4.5 Other small X-ray missions

Ariel 5 was the longest lived of the early X-ray satellites, but it was by no means the last. In late 1974 the USSR launched the Salyut 4 space station which carried two X-ray experiments. 'Filin' was a package of four X-ray detectors plus Sun and star sensors. Three of the detectors were gas filled proportional counters with an effective area of 150 sq cm and a field to view of $3° \times 10°$. They were sensitive to X-rays in the 2–10 keV range. The fourth detector was smaller and operated between 0.2 and 2 keV. The second instrument, called RT-4, was a paraboloid X-ray collector 20 cm in diameter which brought soft X-rays (0.16–0.28 keV) onto a proportional counter detector. The field of view of the instrument was about $1.7°$, but, like the instrument on Copernicus, it was not able to form images; it merely collected X-rays from its entire field of view.

Salyut-4 was initially occupied by Alexi Gubarev and Georgi Grechko who were launched on 10 January 1975 and remained in space for 29.5 days, setting an endurance record for a Soviet spaceflight. The Filin instrument was activated on 15 January and used for observations of Sco X-1 and of the Her X-1 X-ray pulsar. The next visit, launched 5 April 1975, almost ended in catastrophe as cosmonauts Vasily Lazarev and Oleg Makerov were forced to blast their Soyuz spacecraft clear of its launch vehicle as it

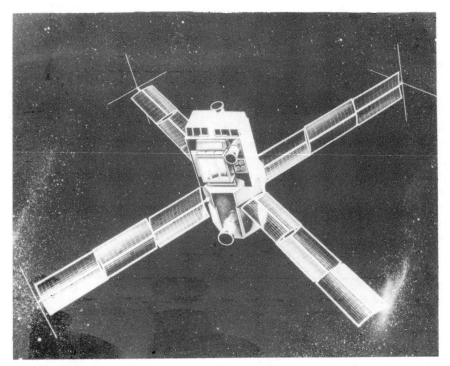

An artist's impression of SAS-3 in orbit. One of the X-ray detectors appears as a square box on the upper portion of the satellite (NASA).

began to tumble out of control 150 km above the Earth. The cosmonauts landed safely and were replaced by Pytor Klimuk and Vitaliy Stevastyanov who were launched on 24 May. During their 63 day mission the Filin and RT-4 instruments were operated on a number of occasions.

An X-ray experiment was also carried on the Indian satellite Aryabhata, named after the 5th century mathematician and astronomer. The satellite was launched by the USSR on 19 April 1975, but a transformer failed soon after reaching orbit and the mission was terminated four days after launch.

More successful was the American Small Astronomy Satellite-3 (Explorer 53) launched from the San Marco platform in 1975. SAS-3 carried X-ray experiments from the Massachusetts Institute of Technology designed to observe discrete extragalactic sources and to monitor the intensity and spectra of galactic X-ray sources in the 0.2–80 keV range. Like SAS-1, SAS-3 was spin stabilised, but its rotation rate was only 0.1 degrees per second. The spin axis could be commanded to 'dither' back and forth across a selected object at a rate of about 0.01 degrees per second for very detailed observations. In particular SAS-3 was able to determine the positions of X-ray sources with much greater accuracy than its predecessors.

SAS-3 astronomers were active in the hunt for X-ray bursters, and their detailed studies contributed greatly to the eventual explanation of the phenomenon. They also discovered the so called 'Rapid Burster', an X-ray star which seems to burst very frequently in a regular fashion and then, after a particularly energetic burst, go quiet for a time. The satellite made a number of important observations of galactic X-ray sources, detecting X-rays from white dwarf stars and from one or two normal, that is, non-collapsed stars. The detection of normal stars suggested that many more such stars might be detectable with a more sensitive X-ray instrument.

Like Ariel 5, SAS-3 observed X-ray sources beyond our Galaxy and confirmed that X-rays were emitted from active galactic nuclei. SAS-3 observations of the unusual galaxy Centauraus A (NGC 5128), known to be both a radio and an X-ray source, were crucial in determining the location of the X-ray emitting region. Results from the SAS-3 rotation modulation collimator showed that the X-rays came from a point source less than 10 arc seconds from the centre of the galaxy, unlike the radio emission which originates both in the centre of the galaxy and in two enormous extended regions on either side.

A package of X-ray astronomy instruments was carried in the wheel section of the final Orbiting Solar Observatory, OSO-8 (launched 21 June 1975). One objective of these experiments was to study the diffuse X-ray background discovered by earlier experiments. At high energies (that is, greater than about 1 keV) the X-ray background is uniform over the entire sky, but at lower energies the structure is more complex. The uniformity of the high energy background implies that it arises beyond our Galaxy (otherwise the emission would be concentrated along the Milky Way), and astronomers using data collected as OSO-8 swept regularly across the sky hoped to gain a clearer understanding of whether the background was genuinely diffuse or if it originated in many thousands of distant X-ray galaxies. Equally of interest was the reason for the complex structure of the low energy X-ray background; was it due to clumping of the soft X-ray sources, or to absorption of the X-rays from distant objects by clouds of nearby interstellar gas? OSO-8 also carried an X-ray spectrometer to study the temperatures, ionisation states, and abundances of material in X-ray emitting regions such as supernova remnants and X-ray binaries, and an instrument to study polarisation in X-ray sources.

OSO-8 operated successfully for over two years, although the details of the X-ray background remained mysterious even after the mission. Among other achievements, the satellite confirmed the presence of emission lines from highly ionised iron in galaxy clusters, identified a number of X-ray binaries, and made further studies of X-ray bursters and supernova remnants.

2.4.6 HEAO-1, end of an era

The era of these small X-ray satellites ended in 1977 with the launch of the first of NASA's High Energy Astrophysical Observatory (HEAO) missions. As noted in Chapter 1, the evolution of the HEAO programme was a protracted process, and the details are described in detail elsewhere (W. Tucker's *The star splitters and The X-ray universe*, and W.&K. Tucker's *The cosmic enquirers—see Bibliography*) and will not be repeated here. It is sufficient to note that the original programme called for four large spacecraft weighing about 12 000 kg each and carrying a variety of very large X-ray, gamma ray and cosmic ray experiments. The programme was cancelled in January 1973, but lobbying by scientists persuaded NASA to change the cancellation to an 18 month 'suspension' and eventually to resurrect the project as three smaller spacecraft weighing about 2700 kg each. To reduce costs, each spacecraft was based around the same equipment module and used a high proportion of 'off the shelf', flight qualified components. Other cost cutting measures included the production of only one model of each HEAO (the protoflight approach) and designing each satellite for a limited lifetime, restricting the ultimate cost of mission operations.

Under the restructured HEAO programme the first satellite would undertake a six month mission to survey and map the entire sky over a wide range of X-ray energies. This was similar to the mission of Uhuru (an early proposal referred to the mission which became HEAO-1 as a 'Super Explorer'), but using a larger satellite equipped with more sensitive instruments able to determine the spectra of X-ray sources and to monitor the diffuse X-ray background in the 0.15–60 keV range.

HEAO 1 was a hexagonal structure 6.1 m long and 2.35 m in diameter. Approximately half of the length was accounted for by the equipment module, which had a cylindrical tank down the centre containing attitude control propellant. Electronic components were attached around the outside of the central structure and were accessible from the outside via removable panels. These data handling system could store about 1.72 megabits of data, enough to record data for two orbits if a regular

The HEAO-1 satellite undergoes pre-launch tests. (NASA).

ground station contact was missed. Attitude control was provided by gyroscopes and gas jets.

The largest experiment was the Naval Research Laboratory Large Area X-ray Survey Experiment, a modular assembly of seven proportional counters, six on one face of the spacecraft and one on the opposite side, sensitive to X-rays in the 0.15–20 keV range. The total area of these counters was 14 000 sq cm, and, depending on their operating mode, they could record signals with a time resolution of up to 5 microseconds. This very high resolution was possible only when the satellite was passing over a ground station and data could be transmitted directly to the ground.

The Cosmic X-ray Experiment used six collimated proportional counters to measure the emission and absorption of diffuse X-rays in three different energy bands (2–60 keV, 1.5–20 keV, and 0.2–3 keV). Four of the detectors viewed the same region of sky; the other two were offset by 6° to allow the background radiation to be monitored. It was planned to compare the data on diffuse emission with observations from radio and optical telescopes, and a major objective was to search for unevenness in the X-ray background which might provide a clue to its origin.

The Scanning Modulation Collimator Experiment was used to measure the position of X-ray sources with an accuracy of a few arc seconds. The experiment consisted of two collimators inclined at $+10°$ and $-10°$ to the scanning directions, allowing the location and structure of sources to be determined along two directions. Star trackers were included in the experiment to assist in locating the X-ray sources relative to ordinary stars.

Finally, the Hard X-ray and Low Energy Gamma Ray Experiment used scintillation counters to detect photons between 10 keV and 10 MeV. It was capable of providing information on the position, spectrum and variability of hard X-ray sources and was able to monitor the spectrum and isotropy of the gamma ray background.

HEAO-1 was launched by an Atlas Centaur rocket on 12 August 1977 and, as planned, the first six months of the mission were spent on an all sky survey. When this was complete a second survey was begun, together with a limited number of pointed observations. The ability of HEAO-1 to make pointed observations was not required for the basic mission, but was included as a relatively cheap way of increasing the scientific usefulness of the satellite.

Despite the planned lifetime of one year, HEAO-1 continued to operate until the second week of January 1979 when its attitude control gas was depleted. The satellite burned up during re-entry later in the year. Throughout its lifetime HEAO-1 used its superior sensitivity to observe faint sources and to locate them more accurately than any previous satellite. The result was an X-ray catalogue more complete than any before it. HEAO-1 observed numerous X-ray binaries and detected X-rays from many more normal stars than had been possible with its smaller predecessors. The spectra of many galaxies and galaxy clusters were also recorded.

Mapping of the diffuse X-ray sky also produced a major new discovery, the existence of a bubble of hot gas some 1000 light years across. The Cygnus Superbubble, as it became known, was at first thought to be a supernova remnant, but subsequent studies showed that the superbubble was 10 times larger (and hence has 1000 times the volume) than a normal supernova remnant. It also contains 20 times more energy than expected. Astronomers think that the bubble was blown by a series of supernovae, each of which sent shock waves into the local interstellar gas. These shock waves pumped more energy into the superbubble and led to the production of new, massive stars which themselves soon became supernovae. The superbubble thus resulted from that most spectacular of events, a chain reaction of supernovae. If this explanation is correct then the interstellar gas throughout our galaxy is likely to be laced with low density tubes and tunnels formed where superbubbles have merged into one another.

Despite its success, HEAO-1 marked the end of an era in X-ray astronomy. Its huge detectors were only about seven times as sensitive as the Uhuru satellite, because the sensitivity of large area proportional counters increases only as the square root of the detector area. HEAO-1 marked the sensible upper limit for large area counters; to produce an instrument with improved sensitivity a new technique would be required. Fortunately, just such an instrument had been under development for some years; even as HEAO-1 was reaching the end of its life, the first X-ray telescope designed to image objects beyond the solar system was launched aboard HEAO-2.

2.5 THE X-RAY OBSERVATORIES

2.5.1 HEAO-2 (Einstein)

A telescope that could focus X-rays, rather than simply collect them, had been an objective of Giacconi's group at American Science and Engineering ever since they had begun their X-ray astronomy programme. After the success of the Uhuru satellite Giacconi became more determined than ever to produce the World's first large X-ray telescope.

X-rays cannot be focused by normal optics, so an X-ray telescope must operate by using grazing incidence reflection, which occurs when X-rays strike a polished metal surface at a very small angle and are reflected rather like a flat stone bouncing off the surface of a pond. The physics of grazing incidence reflection had been established in the 1930s when it was realised that, at X-ray wavelengths, the index of refraction in some materials is smaller than it is in vacuum. This means that X-rays striking a surface at suitably small angles will not penetrate the surface, but will instead undergo total external reflection. The efficiency of this process is determined by the reflecting surface. Metals with low atomic numbers are highly reflective (with an efficiency of about 90%) to X-rays arriving at an angle of less than about 1°, but their reflectivity drops dramatically at larger angles. Higher atomic number metals are less efficient reflectors for X-rays incident at less than 1°, but remain effective at angles up to about 2.5°. In the 1950s the German physicist Hans Wolter suggested a number of reflection systems which might be used to make an X-ray microscope, but at the time there was no technology capable of polishing the mirrors required for such an instrument. The larger mirrors required for an X-ray telescope were, however, within reach of the early X-ray astronomers and Wolter's designs were adapted for use in space.

The small angles required for grazing incidence reflection mean that X-ray telescopes bear no resemblance to their optical counterparts. They are deep paraboloids or hyperboloids of revolution, with a reflective metal coating on their inner faces. A crude, although reasonably accurate, description of such a system is that it resembles a bottomless bucket with a highly reflective metal plated interior. X-rays entering the top end of the bucket make glancing reflections from the inside walls and come to a focus somewhere beyond the bottom.

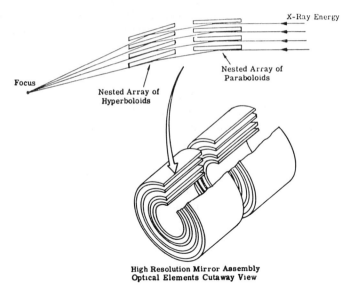

Schematic diagram of a grazing incidence X-ray telescope as used in the HEAO-2 (Einstein) observatory. (Perkin-Elmer Corporation)

The first X-ray telescope built by Giacconi's group at AS&E was used to obtain X-ray images of the Sun from a rocket in 1963, but Giacconi really wanted a telescope for extra-solar astronomy. Unfortunately he was trapped by a Catch-22; the considerable funds needed to develop a large X-ray telescope would be available only if a mission requiring one was approved, but no mission would be approved until the concept of an X-ray telescope had been proved. Giacconi's solution was to persuade the OSO project to include an X-ray telescope as an experiment on the planned Advanced OSO satellite. The Advanced OSO was abandoned in 1965, but some of its experiments (including two X-ray telescopes) were eventually included in the Apollo Telescope Mount, a package of solar instruments launched as part of the Skylab space station in 1973. The Skylab instruments returned stunning X-ray images of the Sun, and proved the feasibility of a large X-ray telescope.

About the same time that the Advanced OSO was cancelled there was a meeting of American astronomers at Wood's Hole, Massachussets from which the original proposals for the HEAO missions emerged. One of the recommendations of the meeting was the development of a large X-ray telescope for extra-solar astronomy, and, despite the changes to the HEAO programme over the next decade, the X-ray telescope survived to become the second mission in the series, HEAO-2.

The HEAO-2 telescope consisted of four paraboloid–hyperboloid mirrors one inside each other, and it was able to focus X-rays with energies between about 0.25 and 4 keV. This complex arrangement of mirrors was determined by two considerations; the desire to form images with comparable angular resolution to those from optical telescopes, and the need to make the instrument as sensitive as possible. A simple paraboloid, such as the one used in the X-ray telescope on Copernicus (see section 2.4.3) can collect X-rays from a small region of sky, but there are severe aberrations produced by a single reflection telescope which make it impossible to form an image. Only if the paraboloid is followed by a hyperboloid, forcing the X-rays to undergo two glancing reflections,

The X-ray telescope for the HEAO-2 (Einstein) observatory being prepared for testing in a specially built facility at the George C. Marshall Spaceflight Centre. (NASA).

can the photons be brought to a focus with sufficiently small aberrations to produce an image with an angular resolution of an arc second or so. The HEAO-2 telescope used a nest of four mirrors to increase the effective collecting area, and so improve sensitivity. Nesting of mirrors in an X-ray telescope is possible, indeed desirable, because only photons which are near to the edges can undergo grazing incidence reflections; photons which enter near the centre are not reflected. Since the centre volume of the instrument is effectively wasted, it is sensible to fill it with a smaller mirror designed to focus X-rays onto the same point as the outer mirror. The central volume of this smaller mirror is also wasted, so a yet smaller mirror is placed here, and so on. With suitably precise engineering, the nested mirrors can be co-aligned to produce a single image, making the telescope much more sensitive than one with just a single mirror.

Since no X-ray mirrors this large had ever been made, they were critical to the development of HEAO-2, and production was placed in the hands of the Perkin-Elmer corporation, a company with considerable experience in producing mirrors for space applications. First, glass mirrors were ground into the correct shape, using diamond tipped tools. Then they were polished and thin layers of chromium and nickel were coated onto the glass to achieve the required reflectivity. Finally, the mirrors were aligned and glued to a support structure, with due account being given to the fact that they were assembled under gravity, but would be weightless once in space. The complete mirror assembly was then taken to the Marshall Spaceflight Center and tested in a specially built facility. The telescope field of view varied with the energy of the incoming X-rays, but was about 1 square degree. At the focal plane was a turntable containing four scientific instruments of different types. By rotating the table a different instrument could be brought into use as required.

To take maximum advantage of the angular resolution of the telescope, HEAO-2 carried a High Resolution Imager (HRI), an instrument which consisted of two microchannel plates (placed one above the other) with a position sensitive detector underneath them. A microchannel plate is a parallel array of tiny glass tubes each coated with a layer of material that emits an electron when struck by an X-ray. A potential is applied along the length of the tubes so that when an X-ray enters one of the tubes and produces a photo-electron, the electron is accelerated down the tube and liberates more electrons as it bounces off the walls. In the HRI, the electron cascade that emerged from the microchannel plate fell onto a fine grid of parallel wires (about 50 wires per centimetre with equally spaced gaps between them), and electronics mounted around the edge of the grid sensed the arrival of the electron cloud. Provided that the X-rays did not arrive too quickly (more than about 100 per second) it was then possible to calculate the position on the grid at which the centre of each electron cloud arrived, and, once the arrival positions of many photons had been determined, an X-ray image could be formed by computer processing. Although capable of arc second resolution, the HRI had a field of view less than half a degree in diameter and so used only part of the field visible to the telescope. Furthermore, since all the information about the energy of the incoming photons was lost during the detection process, the HRI provided no spectral information.

Spectral as well as positional information was provided by the Imaging Proportional Counter (IPC). The IPC was a proportional counter in which imaging was accomplished by two planes of cathode wires, one on either side of the anode wires across the gas filled chamber. An X-ray entering the detector would set off a charge avalanche proportional to the energy of the photon in the usual manner, but the electron avalanche also induced signals in the cathode wires from which the arrival position of the photon was extracted electronically. The spatial resolution of the IPC was poorer than that of the HRI (about 1 arc minute), but it covered the entire field of view of the telescope and was more efficient than the HRI. The two instruments were complementary.

For cases where more detailed energy resolution was required HEAO-2 carried two spectrometers. The cooled Solid State Spectrometer was developed by the Goddard Space Flight Center and provided an energy resolution of about 0.15 keV. The Bragg curved crystal spectrometer from the Massachusetts Institute of Technology had better resolution, about 0.001 keV, but was rather less sensitive. Another source of spectral information was an objective grating spectrometer which could be used in conjunction with the HRI or IPC. A fifth instrument, the Monitor Proportional Counter (MPC) was mounted piggy back on the telescope and studied a wider band of X-ray energies. The MPC was used to examine the spectra and variability of the brightest sources in its field of view.

HEAO-2 was launched on 13 November 1978. Four days after launch, and after some problems with the spacecraft's star trackers had been resolved, the observatory was pointed at its first target, the black hole candidate Cygnus X-1. Good signals were detected and an image of the source was quickly built up by the IPC. The first true orbiting X-ray telescope was working, and working well. Soon after launch the satellite was named the Einstein Observatory, to celebrate the centenary of the famous physicist's birth. The name was chosen by a vote amongst members of HEAO-2's scientific team. Like the naming of Uhuru, the name was not formally approved by NASA, who still refer to the satellite as HEAO-2, and like Uhuru, the name has stuck and is now widely used throughout the scientific community.

The achievements of astronomers using the Einstein Observatory are too great to detail here, and further reading can be found in the *Bibliography*. The most important features of the satellite were its ability to detect faint X-ray sources, and to make images of complex or extended sources. These capabilities made Einstein a revolutionary advance; it brought X-ray astronomy into the mainstream of astronomical research by showing that almost every type of astronomical object was a source of X-rays.

For example, Einstein observations of normal stars provided a number of surprises showing that the X-ray flux from young massive stars (spectral types O, B, and A), as well as from the fainter K and M stars, was thousands of times greater than predicted. These observations demonstrated that existing models of stellar X-ray production were in need of considerable overhaul.

Detailed observations were made of several supernova remnants with studies of the Crab Nebula showing that it produces X-rays by synchrotron emission, as high energy electrons are accelerated by the magnetic field of the spinning neutron star at the centre.

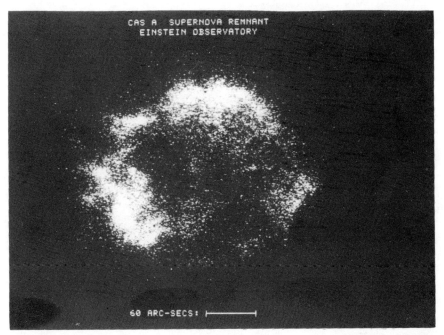

This image of the supernova remnant Cassiopea A was made by the HEAO-2 (Einstein) Observatory. (NASA).

Older remnants, such as Cassiopeia A and Tycho's supernova, were imaged enabling the complex shells of X-ray emitting gas to be mapped in detail. The spectra and temperature of the gas in the shells were determined, and from these, limits were placed on the chemical composition of the supernovae progenitors. Observations of nearby galaxies detected supernova remnants in the Large Magellanic Cloud (LMC). By combining these measurements with optical and radio observations astronomers were able to estimate the ages (1000–20 000 years) and initial energies (about 10^{43} joules) of these supernovae and to calculate the density of the interstellar medium in the LMC. The values derived suggest that the interstellar medium of the LMC is not a simple homogeneous gas, but includes dense clouds embedded in a low density medium.

Einstein observations also confirmed the existence of X-ray binaries in the Andromeda galaxy. By joining a number of images together it was possible to show that most of the sources are found in the spiral arms, although there is also a sharp concentration at the very centre of the galaxy. It was also found that the central sources are, on average, twice as luminous as the ones in the spiral arms. It is worth noting that previous experiments had only just been able to detect the Andromeda galaxy, and could provide no information on the location or the number of individual X-ray sources. The existence of a concentrated region of X-ray emission at the centre of the radio galaxy Centaurus A was confirmed, and X-ray emission lying along the direction of its optical jet was detected. The X-rays seem to arise from Bremsstrahlung in a hot plasma and may provide a clue as to how energy is transported into the galaxy's enormous radio lobes. X-rays were also detected from a variety of other active galaxies, and in some cases the discovery of short term variability revealed that the radiation must come from a central region only a few light-hours in diameter.

Observations of clusters of galaxies by other satellites had shown that X-ray emission came from a 10–100 million degree gas spread throughout each cluster. Einstein observations showed that the regions around individual galaxies are also sometimes sources of X-ray emission. This emission may be produced by gas lost from stars within a galaxy which is heated by supernovae and escapes from the gravitational well of the galaxy into intergalactic space. In some cases the gas may be confined close to the galaxy by the pressure of the still hotter gas between them; in others it may be stripped away as each galaxy orbits around the centre of mass of the cluster and passes through the denser central regions of intracluster gas. Studies of the distribution of this hot gas are providing important clues on the evolution of clusters of galaxies. Observing even deeper into space, Einstein detected X-rays from distant quasars. Long observations of apparently empty fields of view revealed new faint quasars and showed that much, if not all, of the diffuse X-ray background may be due to the combined emission from many faint sources.

The Einstein mission ended in April 1981 when the satellite ran out of gas for its attitude control thrusters. Pointing of the spacecraft was accomplished by momentum wheels, but thrusters were used to unload angular momentum from the wheels. The scientists had pressed for a magnetorquer system to unload momentum, but the decision to use gas jets had been made in order to limit the mission lifetime as part of the strict cost control of the revised HEAO programme. Despite this, mission controllers worked out complex pointing schedules to minimise the amount of momentum built up by the spacecraft, and succeeded in making the gas supply last more than twice as long as originally planned. The satellite re-entered and was destroyed on 25 March 1982.

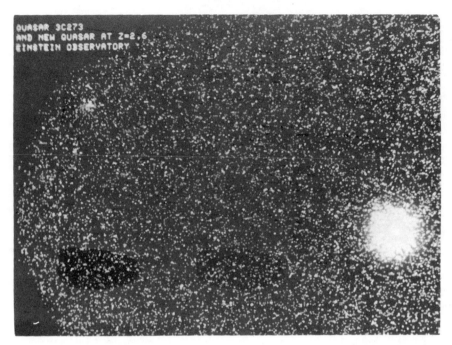

The quasar 3C273 (bottom right) and another, newly discovered quasar (top left) imaged by the HEAO-2 (Einstein) telescope. NASA.

2.5.2 Small observatories

While Einstein was still operating, two other X-ray satellites were launched. First of these was the Japanese Hakucho satellite, orbited in February 1979. Hakucho, known before launch as Corsa-B, made observations of X-ray bursters and soft X-ray sources and began a series of increasingly sophisticated Japanese X-ray astronomy satellites.

Less successful was the British Ariel 6 mission. Launched in June 1979 it was the last in the Ariel series and the only one which was not part of a joint programme with the USA. Although launched by an American Scout rocket, the launch was bought from NASA and not provided free under an international programme. The 150 kg, spin stabilised, satellite was basically cylindrical in shape, with a spherical compartment at one end containing a large cosmic ray experiment. Four solar panels projected at right angles to the cylindrical body. Ariel 6 carried two small X-ray experiments. The first of these used grazing incidence reflectors and thin window proportional counters to observe photons with energies below 2 keV; the second was used to monitor variable X-ray sources in the 1–50 keV range. High time resolution was the objective of the latter experiment which had a 3° field of view defined by collimators. Both X-ray experiments viewed along the spin axis.

Unfortunately, Ariel 6 did not repeat the success of its predecessor. There were problems with the spacecraft's command encoders, and both the tape recorders and batteries proved unreliable. These difficulties severely curtailed results from the X-ray instruments, and the satellite was deactivated in February 1982.

A second Japanese X-ray satellite was launched on 20 February 1983. Astro-B,

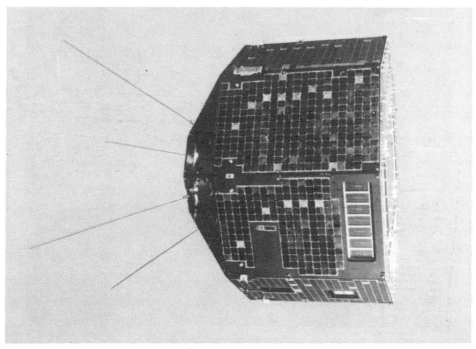

The Japanese Hakucho satellite. (ISAS).

named Tenma (Pegasus) after launch, was dedicated to providing improved spectral and temporal resolution of X-ray sources. Its main instrument was an array of 10 Gas Scintillation Proportional Counters (GSPCs) operating in the 2–60 keV range. A GSPC converts incoming X-rays to photoelectrons and accelerates them by means of electric fields until they have sufficient energy to excite gas atoms in the detector. The excited gas atoms combine to form molecules which then de-excite by emitting ultraviolet photons. The photons are detected by a photomultiplier, the output of which is proportional to the energy of the original X-ray. Although more complicated in design, the GSPC has better energy resolution than a conventional proportional counter. A second experiment used a pair of one-dimensional focusing mirrors and Position Sensitive Proportional Counters (PSPC) to detect X-rays in the 0.1–2.0 keV range.

Tenma was also equipped with a Transient Source Monitor, an experiment consisting of two detector groups. The Hadamard X-ray telescope used a PSPC in conjunction with a mask which cast a shadow onto the detector. A one-dimensional X-ray image could be constructed from this shadow, and the use of two instruments with mask patterns aligned at right angles allowed the position of a single bright source in the 40 degree square field of view to be defined. If several bright sources were detected simultaneously the analysis was more complex, and it was necessary to combine data from each telescope with knowledge of the spacecraft's spin to produce a two-dimensional image. The second part of the instrument used a pair of collimated proportional counters with a field of view covering about 2×25 degrees each.

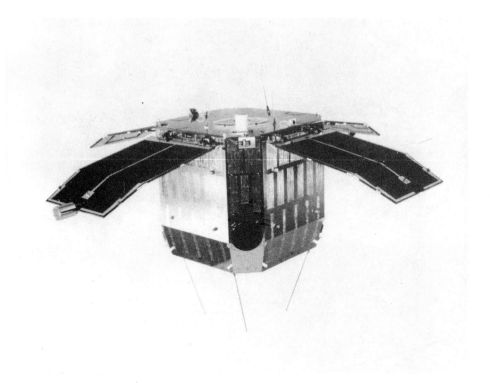

The Japanese Tenma satellite. (ISAS).

2.5.3 EXOSAT

The next major X-ray observatory was the European EXOSAT, a satellite with a long and complex history. The mission was originally proposed in 1969 as HELOS (Highly Eccentric Lunar Occultation Satellite), a 150 kg spacecraft designed to locate X-ray sources by determining the instant they disappeared behind the Moon. In 1973 the HELOS mission was expanded to include a limited capability for pointed observations and was renamed EXOSAT (European X-ray Observatory SATellite). As a consequence of the additional objectives, the mass of the satellite grew to 300 kg.

Despite the 1973 go-ahead, financial restrictions delayed the main design phase (Phase B) until 1977, by which time the results from Uhuru and Ariel 5, together with the forthcoming launch of the Einstein Observatory, had made it obvious that EXOSAT was being overtaken by events. At the cost of an additional launch delay, it was decided that the mission should be restructured into a full X-ray observatory. At about the same time the planned launch vehicle was changed from an American Delta to a European Ariane rocket. As a result the mass increased again, first to 480 kg and then to 500 kg.

Although EXOSAT was now planned for an Ariane launch, it was required to retain compatibility with the Delta rocket in case of delays in the Ariane programme, a fortunate decision since the original launch was cancelled to make way for a commercial customer after the failure of the fifth Ariane. EXOSAT was eventually launched by a Delta in May 1983 and was placed into a highly inclined orbit with a

The EXOSAT satellite before launch. (ESA).

perigee of 350 km and an apogee of about 190 000 km. This orbit was chosen so that large areas of the sky would be occulted by the Moon throughout the mission, a relic of the original HELOS proposal, but which was equally suitable for an observatory mission.

EXOSAT was built by a consortium of European aerospace contractors led by the German MBB company. It was a box-like structure, 2.1 m in diameter and 1.35 m high, topped by a solar panel which could be rotated to face the Sun irrespective of the direction in which the satellite was observing. Attitude control was provided by propane gas thrusters, and a motor using catalytically decomposed hydrazine as propellant was used to modify the satellite orbit so that sources could be occulted by the Moon as required. EXOSAT carried three experiments, each built by different international teams. All three experiments looked out of the same side of the spacecraft.

The large area proportional counter, otherwise known as the Medium Energy (ME) experiment, was a survivor from the original HELOS concept. It consisted of an array of eight multiwire proportional counters with a total effective collecting area of about

Impression of EXOSAT in orbit. Note the two telescope apertures (circles) and the proportion counter arrays which are revealed when the covers open. (ESA).

1800 sq. cm. Each counter had a front section containing an argon/CO_2 mixture (energy range 1.5–15 keV, but optimised for 2–6 keV) and a rear element containing xenon and CO_2 (energy range 5–50 keV). The time resolution possible with the instrument was about 10 microseconds, making it suitable for studies of pulsars and other rapidly varying sources. The ME had two interesting design features. The first of these was the collimator, which needed to define a narrow field of view yet fit into a limited space. The solution was a lead glass collimator based on methods used in microchannel plate production. Starting with a cylindrical tube of lead glass, a series of drawing, cutting, and fusing operations were undertaken to produce a collimator only 11 mm high and with a field of view of only 45 arc seconds. The second feature was a mechanism to allow some elements of the array to be tilted until they observed about 2° off the main viewing axis. This allowed the background detection rate to be monitored simultaneously with observations of a selected source.

The second experiment consisted of two identical telescopes capable of imaging at energies up to about 2 keV. Each telescope used two nested paraboloidal/hyperboloidal mirrors in a Wolter type 1 configuration and was limited to 1 m focal length by the size of the satellite. Each telescope was equipped with two detectors on a rotating plate, either of which could be used at any time. One detector was a chevron arrangement of two microchannel plates known as a Channel Multiplier Array (CMA); the second was a position sensitive proportional counter called the Position Sensitive Detector (PSD). To avoid contamination of its observations by ultraviolet light, the CMA was always operated in conjunction with a UV blocking filter. Like the Einstein HRI and IPC, the CMA had the better angular resolution at the cost of spectral information, while the PSDs were capable of providing spectral information but had lower spatial resolution. Both telescopes were equipped with a diffraction grating, to be used in conjunction with the CMA for high resolution spectroscopy.

The third experiment was a gas scintillation proportional counter operating between 2 and 80 keV. This provided better energy resolution than the ME, and was used to probe the spectra of bright X-ray sources.

EXOSAT suffered several problems soon after launch. Both PSDs failed, the CMA in one telescope stopped operating, and a mechanism used to position the diffraction grating in the field of the other telescope jammed. This was the worst combination of failures possible, and it severely limited the amount of spectroscopy possible at low energies. Various anomalies also occurred in the attitude control system which, although eventually resolved, caused a considerable drain on the limited propane gas supply.

Some of these difficulties were rectified (but not the problems with the X-ray telescopes), and the satellite controllers eventually settled down to routine operations. EXOSAT's orbit allowed long and uninterrupted observations of individual sources and permitted long periods of real time operations. This enabled EXOSAT to make coordinated observations with other satellites or with ground based telescopes, and about 25% of observations were scheduled in this way. Only a few scientific highlights will be mentioned here; others can be found in specialist journals, and many more have yet to be published since data analysis is still going on.

EXOSAT was used to study numerous binary systems including monitoring a complete cycle of the variable AR Lac, a system containing two solar type stars orbiting very close together. These stars emit X-rays from their coronae and are classified as RS CVn stars. Their activity is believed to be related to massive starspots on one or both stars which release energy into the coronae in some way. Many other X-ray binaries were studied, and one was found with an 11-minute orbital period. This is so short that the only likely explanation is that the X-rays are emitted during mass exchange in a white dwarf/neutron star system; only two collapsed stars could get close enough together to have such a short orbital period. Another discovery was the phenomenon of quasi periodic oscillations, very rapid X-ray variability observed in a group of sources within our Galaxy. The origin of these rapid variations remains unclear, but may be related to clumps of material orbiting close to a neutron star. EXOSAT also observed active galaxies, and its detection of short-term variability in the galaxy Markarian 766 implies that its central X-ray source can be no larger than 1000 times the diameter of the Sun. The most likely explanation is that the centre of Markarian 766 contains a supermassive black hole.

EXOSAT was so successful that when the planned two year mission ended, approval was given for an extension until the satellite reached the end of its useful life. This came on 9 April 1986, a few months earlier than expected, after a failure of the attitude control system led to the premature depletion of the remaining control gas. EXOSAT re-entered the atmosphere and was destroyed on 6 May 1986.

2.5.4 X-ray imaging at high energies

Grazing incidence optics are effective only at energies below about 4 keV; at higher energies the angle of incidence required for grazing reflection is so small that a telescope with an adequate collecting area would be impracticably large to manufacture. Although images can be built up by scanning a detector across a source, this is very inefficient because only a short time can be spent observing each picture element, and the image may be degraded if the source varies during the observation. The first orbiting instrument able to make images directly at high energies was a new type of X-ray Telescope (XRT) flown on the Spacelab 2 mission in 1985. The Spacelab XRT, developed by the University of Birmingham, used a coded mask technique to make images at energies of 2.5–25 keV.

A coded mask telescope works rather like a pinhole camera; photons pass through holes in a mask and are registered on a detector (in the case of a camera the detector is a piece of photographic film). A pinhole camera is, in effect, a coded mask telescope with only one hole, and it produces a single image. A camera with two pinholes close to each other would produce two overlapping images, one with three pinholes would produce a triple exposure, and so on. Such multiple exposures would be useless for ordinary photography, but in X-ray astronomy, where position sensitive proportion counters can measure the location and time of arrival of individual photons, it is possible to unscramble a complex multiple image by using a computer. The advantage of using a mask with many holes is that more photons are detected in a given time, an improvement known as the multiplex advantage. A coded mask telescope uses a complex pattern of holes to cast a 'shadowgram' of a mask onto a detector. The shadowgram does not bear any obvious relationship to the object being observed, but by using the position and arrival time of each detected photon together with the pointing history of the telescope and the characteristics of the mask pattern, it is possible to deconvolve an X-ray shadowgram into a detailed image of the sky. If there is only one point source in the field of view, the shadowgram is fairly straightforward—it is a shadow of the mask. However, if there are two or more objects visible, each source casts its own shadow, and the result is a more complex pattern.

To minimise the data processing required the mask pattern is designed to make it easy to find the shadow of the mask even when it is shifted about on the detector and superimposed on itself many times. Finding the correct pattern must also be possible in the presence of background noise. A regular pattern is unsuitable since it would give virtually the same shadow when viewed from a number of different directions; what is needed is a pattern that fits well when correctly aligned, but fits equally badly however it is misaligned. Various mask designs are possible, but the best ones use two-dimensional cyclic patterns known as psuedo-random masks (see the *Bibliography* for details of coded mask theory).

The principle of a coded mask telescope was proved with a small experiment launched on a Skylark sounding rocket in 1976. The rocket spent about 6 minutes above the atmosphere, long enough for an X-ray image of the galactic centre to be

The Spacelab-2 X-ray telescope. The coded masks are at the top, the position sensitive detectors at the base of each section of tubular structure. The experiment is mounted on a temporary plinth and has not yet been wrapped in its thermal protection blankets. (SERC).

made. After image reconstruction a number of new hard X-ray sources grouped around the galactic centre were discovered. The success of this flight led to an invitation to fly a larger coded mask telescope on the Space Shuttle as part of the Spacelab programme. The Spacelab instrument consisted of two co-aligned telescopes mounted together on an alt-azimuth system carried in the payload bay of the Space Shuttle. The use of an separate mounting under the control of a dedicated XRT computer made the instrument partly independent of the pointing of the Space Shuttle and provided greater freedom in target selection (although manoeuvring room was limited to avoid the risk of colliding with other instruments).

The main elements of each telescope were a gold coated mask, in which opaque elements occupied about half the area of the mask, supported about 3 m from a position sensitive proportional counter. The two masks used different sized holes, producing different angular resolutions (3 arc minutes and 12 arc minutes) on the sky. The high resolution telescope was for detailed studies of the brighter sources, the other for observations of regions of fainter, diffuse emission. Each telescope had a field of view

The TTM/COMIS telescope. Note the star tracker camera above the main telescope. The coded mask is behind the protective black aperture cover.

about 6° wide. A TV system and an ordinary photographic camera viewed along the telescope axis to provide extra pointing information which could be combined with attitude data from the Space Shuttle computers to assist in image reconstruction after the mission.

The first flight of the XRT was on the Spacelab-2 mission in July and August 1985. The Shuttle failed to reach its intended orbit, because of the premature shutdown of one main engine, but this did not affect the XRT which functioned extremely well throughout the flight. Over 75 hours of high quality data were obtained, and observations were made of eight clusters of galaxies, the central region of our own Galaxy, and of the Vela supernova remnant. Like all Spacelab hardware the XRT is designed to be refurbished and relaunched, and may be flown again in the future.

2.5.5 The Kvant module

The Kvant (quantum) module is an element of the Soviet Mir space station complex and contains an international package of experiments for high energy astronomy known collectively as the Rentgen observatory. The module was launched in March 1987 and docked to the space station a few days later. One of the experiments is a coded mask telescope, known variously as COMIS or TTM, which was developed jointly by the University of Birmingham in the UK and groups in the Netherlands. The COMIS telescope has a field of view 15° square, and will be used to map a number of complex regions of X-ray emission in the energy range 2–30 keV. In addition to the COMIS telescope, Kvant carries three other X-ray experiments. Sirene-2 is a gas scintillation proportional counter provided by ESA. The instrument is based on the EXOSAT GSPC (see section 2.5.3) and operates over the 2–100 keV range. A scintillation spectrometer, known as HEXE, has been provided by a West German team from the

Max Plank Institute for Extraterrestrial Physics and the University of Tübingen. It operates in the 15–250 keV range and has a field of view about 1.5 degrees on a side. Finally, the Soviet Pulsar X-1 experiment, which uses a crystal scintillator 200 mm in diameter and 30 mm long, is able to detect hard X-ray and gamma ray sources up to 1300 keV.

The Kvant experiments all look along the same direction and are pointed by the crew of the Mir who aim the entire space station as required, using a set of momentum wheels installed inside the Kvant module. Operation of the instruments is carried out by ground controllers, usually during periods when Mir passes over the equator to minimise the effects of the Earth's radiation belts. During the first six months of operation the Rentgen observatory made over 300 observations, no fewer than 115 of which were to monitor Supernova 1987A in the Large Magellanic Cloud. Other objects studied during this period were the X-ray sources Cygnus X-1 and Hercules X-1. Unfortunately technical problems with two of the experiments (TTM/COMIS and Sirene-2) arose during the mission.

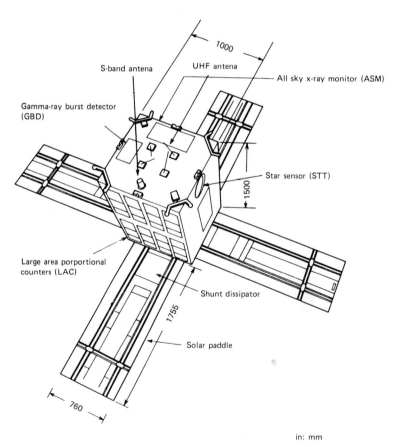

Diagram of the Japanese GINGA satellite showing the location of the scientific experiments. Dimensions are in mm. (ISAS)

2.5.6 Astro-C 'Ginga'

The Japanese Astro-C satellite was launched on 5 February 1987 by a Mu 3S rocket and named Ginga (Galaxy) once in orbit. The spacecraft is box-shaped with a cruciform solar panel array attached to the base. It is three axis stabilised by a momentum wheel and magnetorquers. Ginga is designed to study the time-variability and spectra of X-ray sources.

The main instrument is a large proportional counter array (effective area 4500 cm sq.) developed in conjunction with British astronomers, which operates in the 1.5–30 keV range. The satellite also carries an all sky X-ray monitor, consisting of proportional counters sensitive in the 1.5–30 keV range which is swept across the sky by slow rotations of the spacecraft during intervals when the primary target is eclipsed by the Earth. A gamma-ray burst detector was provided by the Los Alamos National Laboratory of the USA. It consists of a proportional counter and a scintillation counter which between them cover the range 1.5–400 keV.

Ginga was placed into an orbit 505 × 607 km, rather more eccentric than planned, but this has not affected the satellite's mission. The scientific experiments onboard Ginga were activated earlier than planned to observe the supernova in the Large Magellanic Cloud.

2.6 FUTURE X-RAY ASTRONOMY MISSIONS

X-ray experiments are included in the payloads of two Soviet satellites (GRANAT and GAMMA-1); these are described in section 3.7 to place them in the context of the gamma ray instruments which form the main payloads of these two missions. The next

GINGA, (ASTRO-C) (ISAS).

dedicated X-ray astronomy mission will be the German Roentgensatellit (ROSAT), a large satellite due for launch soon after 1990.

ROSAT (named after the German discoverer of X-rays) will carry out the first imaging all sky survey in the 0.1–2 keV range, and will follow this up with an extended period of pointed observations similar to that of Einstein Observatory. ROSAT carries a large, fourfold nested, Wolter type 1 X-ray telescope with a collecting area about three times as great as the Einstein observatory, and it is equipped with two position sensitive proportional counters for the survey phase. These instruments, only one of which is used at a time, have a field of view 2° across and have 1 arc minute resolution. The telescope is also equipped with an American supplied High Resolution Imager (essentially the same as the Einstein instrument) for use during the subsequent pointed phase. ROSAT also carries a British built extreme ultraviolet telescope co-aligned with the main instrument (see section 4.13.2).

ROSAT is a very large satellite (weight over 2 tonnes) originally designed for launch by the Space Shuttle, but, after the loss of the Challenger, it was transferred to an expandable rocket. The satellite consists of a rectangular body containing the X-ray telescope and associated equipment, plus two large deployable solar panels. The extreme ultraviolet telescope is piggy-backed on the telescope compartment, in the shadow of the solar panels. ROSAT uses momentum wheels and magnetorquers for attitude control, and during the survey it will orbit with its solar array facing the Sun and its telescope scanning great circles at right angles to the Sun vector. For the pointed phase, slewing the spacecraft up to 15° off the Sun vector will be allowed.

ROSAT will be placed in a low orbit inclined at about 60° to the Equator and will store about one day's worth of commands in its onboard computer. Data taken during each day's observations will be played back to a German control centre, and new instructions transmitted to ROSAT, during a number of short ground contacts each day. The mission is expected to last 3–5 years.

Other proposed missions include the X-ray Timing Explorer, a small American satellite aimed at observing variable sources with very high time resolution, an Italian X-ray observatory, operating in the 1–200 keV range, called the Satellite for X-ray Astronomy (SAX), and a series of small Space Shuttle payloads known as the Shuttle High Energy Astrophysics Laboratory (SHEAL). Also planned is the American Advanced X-Ray Astrophysics Facility (AXAF), a 10 tonne satellite roughly equivalent to the Hubble Space Telescope (see section 5.1). AXAF, which despite high scientific priority, did not receive funding until 1988, will be 13 m long and will carry a sixfold nested X-ray telescope 1.2 m in diameter and with a 10 m focal length. The satellite will be equipped with a suite of instruments for imaging, X-ray spectroscopy, and polarimetry and is expected to be 100 times more sensitive than the Einstein satellite. AXAF will be carried by the shuttle into a 500 km circular orbit and be operated as a US national X-ray astronomy facility. Periodic visits by astronauts for maintenance and instrument replacement will occur throughout the planned 15 year lifetime of the satellite. A team of eleven scientists who will provide scientific and technical advice during the development of the project were named in May 1985.

Other advanced X-ray missions are under study. The European Space Agency declared an X-ray mission (known variously as the X-ray Multi Mirror Telescope and the High Throughput X-ray Spectroscopy Mission) as one of is main scientific objectives before the end of the century. This is a highly demanding project which requires a large number of grazing incidence telescopes to be nested together. Together with AXAF, this mission will ensure that X-ray astronomy will remain an important element of modern astrophysics well into the next century.

Some highlights of extra-solar X-ray astronomy

1895 Roentgen discovers X-rays during laboratory experiment.
1912 Max von Laue shows that X-rays are photons of electromagnetic radiation which have very short wavelengths.
1938 Theoretical prediction that the solar corona might be a source of X-rays.
1949 NRL group detect X-rays from the Sun with V2 rocket payload.
1952 Wolther proposes designs for X-ray microscopes.
1958 Multiple rocket launches to observe X-rays from Sun during a solar eclipse in an attempt to locate the regions of X-ray emission.
1962 1st extra-solar X-ray source (Sco X-1) discovered during rocket flight.
1963 X-rays detected from Crab Nebula.
      Small grazing incidence telescope carried in a rocket payload takes X-ray images of the Sun.
1964 Lockheed group launch X-ray experiment on rocket with gas jet stabilisation.
1966 X-ray experiment on OAO-1 lost when spacecraft fails.
      First extragalactic X-ray source (the galaxy M87) discovered by rocket.
1967 Launch of small extrasolar X-ray experiment into orbit on OSO-3.
1969 Pulsed X-rays discovered from Crab Nebula.
      Vela 5 satellites used in long-term monitoring of the X-ray sky during a military programme to detect clandestine nuclear tests.
1970 SAS-1 (Uhuru) begins X-ray sky survey.
      US Congress approves HEAO programme.
1972 British X-ray experiment carried on NASA OAO-3 satellite.
1973 HEAO programme cancelled.
      Grazing incidence telescopes on Skylab space station take stunning X-ray images of the Sun.
1974 Reduced HEAO programme restarted.
      Lockheed group detect X-rays from Capella, the first detection of a normal star in X-rays.
      UK Ariel 5 X-ray satellite starts highly successful mission.
      Netherlands/US ANS satellite launched.
1976 X-ray bursters discovered in data from ANS.
      Rapid burster discovered in data from SAS-3.
1977 HEAO-1 launched on sensitive X-ray sky survey.
1978 HEAO-2 (Einstein observatory) carries large X-ray telescope into orbit.
      The telescope detects X-rays from many types of object, and X-ray astronomy enters mainstream of astronomical research.
1983 European EXOSAT and Japanese TENMA X-ray satellites launched.
1985 Spacelab-2 carries large coded mask X-ray telescope on eight-day mission.
1987 Japanese GINGA X-ray satellite and USSR Kvant high energy astrophysics module launched. Both observe Supernova 1987A in Large Magellanic Cloud.
1988 Funds provided for development of Advanced X-ray Astrophysics Facility.

3

Gamma ray astronomy

3.1 INTRODUCTION

The boundary between hard X-rays and gamma rays is, like the others encountered in this book, an arbitrary one. Some astronomers regard gamma rays as being photons with energies extending upward from about 100 keV; others prefer to regard the transition as occurring at energies closer to 500 keV. Wherever the lower energy limit is chosen, there is considerable overlap with the techniques of X-ray astronomy. At very high energies, aspects of gamma ray astronomy become entwined with the study of the energetic charged particles that are known, rather confusingly, as cosmic rays. A few detectors for low energy cosmic rays have been orbited, notably on the HEAO-3 satellite, Spacelab-2, and the Soviet Proton satellites, but since cosmic ray research is a region where astronomy and nuclear physics become difficult to separate, these experiments will not be described here.

Gamma rays are energetic quanta of electromagnetic radiation which can travel great distances through space without appreciable absorption. They are less penetrating than cosmic rays and are absorbed high in the Earth's atmosphere, making them detectable only from high flying balloons. Unfortunately, the sensitivity of balloon borne detectors is reduced by a background of gamma rays produced during the interaction of cosmic rays with the atmosphere, and, to reduce this background radiation, astronomers have taken gamma ray detectors into space. Gamma rays can be produced in a number of different ways, which are summarised below. For more details see the technical works listed in the *Bibliography*.

Low energy gamma rays—those with energies of below a few Megaelectron Volts (MeV)—can be produced thermally, if the source is hot enough (hundreds of millions of degrees), or by the decay of unstable radioactive nuclei. In astrophysical environments the source of these radioactive nuclei is likely to be normal elements which have been transmuted into unstable isotopes by the impact of the high speed particles and high energy quanta found close to energetic objects such as quasars or young supernova remnants.

Above a few MeV there are no known astronomical sources hot enough to produce thermal gamma ray photons, and radioactive decay no longer contributes. The more energetic (that is, with energies greater than about 30 MeV) gamma rays are produced by cosmic rays colliding with atoms in the interstellar medium. In this process a cosmic ray disintegrates a normal atom, producing unstable elementary particles (pions) which decay to form gamma ray photons. Less violent interactions, in which a cosmic ray (usually a high speed electron) passes close by an atom and is curved by the attraction of the atomic nucleus, can produce Bremsstrahlung radiation. Gamma ray photons can also be produced by synchrotron radiation as the paths of cosmic ray particles are curved by powerful magnetic fields, for example near neutron stars.

Another source of gamma rays is the mutual annihilation of matter and anti-matter. This process is also bound up with cosmic rays, for it is cosmic ray interactions with the interstellar medium which produce the anti-particles required. Electron–positron annihilations produce quanta of 0.511 MeV (511 keV), quite a low energy gamma ray; other interactions produce more energetic photons.

Some of these processes, such as nuclear decay and matter/anti-matter annihilation, produce gamma ray spectral lines. If these lines can be detected and identified they can be used to provide information on the nuclei and other particles involved, and can thus shed light on many aspects of high energy physics. The other mechanisms produce continuous spectra, analysis of which reveals the physical conditions in which they are produced.

At very high energies, that is, above 10^{11} eV (100 000 MeV), ground based astronomy becomes possible once again, since these high energy photons interact with the atmosphere to produce showers of elementary particles similar to those produced by cosmic rays. A number of ground based experiments operating in this energy range are producing interesting results, but they fall outside the scope of this book.

3.2 DETECTION METHODS

The very high energy of gamma ray photons means that the only means of detecting them is by their interaction with matter. For photons with energies of below a few MeV, two types of detector have so far been used in satellites: scintillation counters and solid state spectrometers. At higher energies, astronomers must rely on spark chambers similar to those used in nuclear physics research.

All of these types of detector are affected by the background of cosmic rays and charged particles, so, like X-ray detectors, gamma ray instruments include a separate detector sensitive to these particles. If a particle passes through the first detector and also triggers the main gamma ray detector, two signals are seen at virtually the same instant. When such a 'coincidence' occurs the data processing system assumes that a particle, not a gamma ray, is responsible, and it ignores the resulting signal.

3.2.1 Scintillation counters

Scintillation counters have already been mentioned in connection with the detection of hard X-rays. They rely on the fact that low energy gamma rays can interact with matter via Compton scattering, losing energy by ejecting electrons from atoms of a target material as they travel through it. The electrons ejected during the Compton scattering may then interact with the target material to produce tiny flashes of light called scintilla. Photomultipliers, or photodiodes, arranged around the scintillator material measure the intensity of the light flash, and from this the energy of the original gamma ray can be estimated. It is the efficiency and reproducibility with which the photomultipliers collect and amplify the light pulse in the crystal which determine the energy resolution of the detector. The target material, in which the scintillation occurs, can be an inorganic crystal such as sodium or caesium iodide, or a special plastic material containing organic compounds which scintillate in a suitable manner. For space experiments, where detection efficiency rather than cost is the most important consideration, single crystal scintillators are preferred over the cheaper, but less efficient, plastic materials.

The angular resolution of these types of detector is very poor because the direction in which electrons are emitted during Compton scattering is only weakly correlated to the direction of the incident photon. As with the counters used in early X-ray experiments, collimators are used to restrict the angle of entry of the gamma rays, and hence provide improved angular resolution. The collimators may be passive, mechanical, devices which absorb gamma rays before they reach the detector, or may be additional detectors operated as active collimators. An example of an active collimator would be a thin sheet of scintillator material extended parallel to the viewing direction of the instrument. Photons arriving off the viewing axis pass through the collimator material and are registered on their way to the main detector. The experiment electronics, operating in anticoincidence mode, ignore the resulting signal.

3.2.2 Solid state detectors

If high energy resolution is required, that is, if the instrument is to be a gamma ray spectrometer, solid state detectors are required. A common type uses a target material such as silicon or germanium which has been mixed with tiny quantities of other elements such as lithium. The gamma ray passes through the detector, and the energy deposited during Compton scattering causes some of the electrons in the crystal to be raised in energy and allows them to move about in the crystal lattice. Physicists refer to this process as the creation of electron–hole pairs in the material. The freed electrons constitute an electric current which can be collected at electrodes and amplified. The energy resolution of this type of detector is a few keV, a limitation imposed by the high gain, low noise amplifiers required to process the tiny electronic signal deposited in the detector.

One difficulty with this type of detector, which also applies to the scintillation counter, is that when the gamma ray escapes from the target after only one (or occasionally two) Compton scattering the total energy deposited in the crystal is less than the energy of the incident gamma ray, and this reduces the spectroscopic accuracy of the instrument. For this reason germanium is preferred over silicon as a target material for solid state detectors because the higher atomic number of germanium increases its cross-section for photoelectric absorption and increases its efficiency. Both silicon and germanium solid state spectrometers must be operated at low temperatures

Detection methods

to reduce their intrinsic thermal noise, and the more sensitive germanium must be stored at low temperatures even when not in use.

3.2.3 Spark chamber telescopes

For gamma rays with energies greater than a few MeV, the principal interaction with matter is not Compton scattering, but pair production. In this process a photon passing close to an atomic nucleus decays spontaneously into an electron–positron pair. Unlike Compton scattering, the direction of motion of the newly created electron and positron is related to the direction of the original gamma ray, so the path of the incident photon can be reconstructed from the paths of the charged particles produced.

A spark chamber telescope consists of a sealed volume filled with an inert gas (for example a mixture of neon and argon) across which run a series of thin, parallel, metal sheets. When required a high voltage (several kilovolts) can be applied to each alternate plate, while the remaining plates are maintained at the local earth potential. Underneath the stack of plates are one or more triggering detectors, sensitive to the passage of charged particles. When a gamma ray passing through the chamber decays via pair production, the electron–positron pairs travel through the chamber, ionising a trail of gas molecules as they go. Since they have comparatively high energies, the electron and positron are not stopped by the thin metal plates, but continue downwards until they reach the trigger detectors. As soon as the trigger detectors are stimulated, they apply a high voltage between the plates, causing sparks to flash along

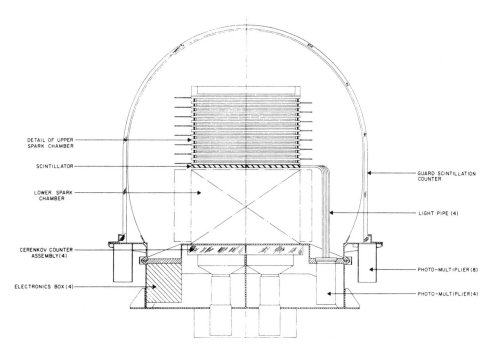

A typical satellite spark chamber experiment. This is the one carried on the SAS-2 mission. (NASA).

the ionised trails in the chamber gas. These sparks mark the path of the electron and positron through the chamber and provide information about the original photon. The sparks may be recorded on film (if the instrument is to be recovered after its mission), acoustically via sensitive microphones, or electronically by thin wire grids stretched across the chamber.

Since spark chambers work by observing charged particles, not gamma ray photons directly, they must be protected against false signals caused by cosmic rays and other charged particles. This is done with an anti-coincidence counter around the detector; charged particles passing through the anti-coincidence counter are detected and the trigger detectors are inhibited, preventing the chamber electrode voltages being applied. Since in Earth orbit charged particles greatly outnumber gamma ray photons, the correct functioning of the anti-coincidence detectors is essential to the successful operation of a spark chamber telescope.

Various refinements of the basic spark chamber instrument are possible. Additional detectors can be fitted below the main chamber and its triggering system to measure the energy of the electron pair as it escapes. Since the total energy of the gamma ray must be conserved during the pair production process, measuring the energy of the electron pair provides information about the energy of the incident photon. The extra detectors may be crystal or plastic scintillation counters and are sometimes referred to as an energy calorimeter. The angular resolution can be improved by subdividing the trigger and additional detectors; knowledge of which regions of these detectors were stimulated provides extra information on the path of the electrons through the chamber and helps to define the direction of the original gamma ray. Despite these additional features, uncertainties in the relationship between the motion of the gamma ray and the electron pair it produces, combined with scattering of the electron pair as the particles travel through the thin metal plates, reduces the angular resolution of such an instrument to a few degrees, which is very crude by normal astronomical standards.

3.3 EARLY EXPERIMENTS

Predictions that extraterrestrial gamma rays would be observable were first made toward the end of the 1950s, but the detection of the predicted flux from balloon borne detectors proved more difficult than expected. The first orbiting experiment was a small scintillation counter carried on Explorer 11 in 1961 (see Table 3.1). The satellite returned data for 62 days, and although it did detect gamma rays, too few were detected to unambiguously confirm that they were of extraterrestrial origin.

A more successful experiment was carried aboard OSO-3 (see section 2.4.1). An MIT scintillation counter mounted in the wheel section of the satellite detected sufficient photons (a total of 631) to establish that there was a broad peak in the intensity of gamma rays in the direction of the galactic centre, and that the total amount of radiation was roughly that predicted by models of cosmic ray interactions with the interstellar gas.

A small, acoustically monitored, spark chamber was carried on the fifth Orbiting Geophysical Observatory (OGO-5). The OGO programme was a series of satellites devoted to geophysical research, and, as with the OSO series, other experiments complementary to the main mission were carried when possible. The OGO-5 gamma

Early experiments

Table 3.1
Extra-solar gamma ray experiments

Name	Launch date	Launch vehicle	Perigee (km)	Orbit[†] Apogee (km)	Inclination°	Notes
Explorer 11	27 Apr 1961	Juno II	485	1616	28.8	Failed 12 June 1961. Marginal detection of gamma rays.
OSO-3	8 Mar 1967	Delta	339	345	32.9	Mostly solar experiments
OGO-5	4 Mar 1968	Atlas Agena D	232	142228	31.1	Small gamma ray Experiment
Cosmos 208	21 Mar 1968	A1/2	129	190	65.0	Military. Re-entry 2 Apr 1968
Cosmos 264	23 Jan 1969	A1/2	136	205	70.5	Military. Re-entry 5 Feb 1969
TD-1A	12 Mar 1972	Delta	531	539	97.5	MIMOSA Expt
SAS-2	15 Nov 1972	Scout	443	632	1.2	Explorer 48
Cosmos 561	25 May 1973	A2	206	295	65.4	Military plus gamma ray experiments
Cosmos 731	21 May 1975	A2	203	296	65.0	
COS-B	9 Aug 1975	Delta 2913	342	99873	90.1	ESA mission, sky survey.
Cosmos 856	22 Sep 1976	A2	212	366	65.0	Military plus 99-2M gamma ray spectrometers operating from 100–4000 MeV. Survey of diffuse gamma ray sky.
Cosmos 914	21 May 1977	A2	210	327	65.0	
Cosmos 1106	12 Jun 1979	A2	214	235	81.4	Crystal scintillation spectrometer.
HEAO-3	20 Sep 1979	Atlas Centaur	424	457	43.6	Cosmic ray experiments and gamma ray spectrometer

[†] Since satellite orbits change because of atmospheric drag etc., orbital parameters quoted by different sources may vary

ray instrument, which was sensitive to photons with energies greater than about 25 MeV, used a six gap spark chamber with an effective area of 102 cm^2. OGO-5 was Earth pointing, and this, together with the need to switch off the instrument when it passed through the Earth's radiation belts, limited the region of the sky which could be observed by the gamma ray experiment to a small area around the direction of the constellation of Cygnus. A variety of problems, including a reduction in the efficiency of the anti-coincidence counter and an anomaly within the data system, severely degraded the results from the experiment, which operated for only five months. During this time gamma rays from the galactic plane were monitored, but no point sources were detected.

The Soviet Cosmos 208, 264, 428, 561, and 731 missions are reported to have carried gamma ray and X-ray detectors. The purpose of the missions was, however, unannounced, and is believed to have been military photographic reconnaissance. The

The Orbiting Geophysical Observatory 5 satellite carried a small gamma ray experiment into a highly elliptical Earth orbit. (NASA).

scientific experiments were carried inside special containers known as Nauka modules. Gamma ray detectors have also been carried on Soviet Prognoz satellites (see section 3.4.2).

The European astronomy satellite TD-1A (see section 4.5) carried a small spark chamber experiment called MIMOSA for observations of gamma rays with energies above 30 MeV. The instrument was equipped with a stereoscopic TV system, viewing through portholes in the chamber, to record the particle tracks. The experiment weighed 31 kg and operated from March to October 1972. Numerous gamma rays were detected, but unfortunately the experiment was affected by particle induced background despite its anti-coincidence system. No point sources of gamma radiation could be identified.

3.4 GAMMA RAY BURSTERS

3.4.1 The Vela programme

The Partial Test Ban Treaty of 1963 forbade the detonation of nuclear weapons in the oceans, the atmosphere, and in space, and, as a consequence, the USA set up a system to monitor clandestine nuclear testing. The programme, code named Vela Hotel, was conceived between 1959 and 1962 and used a series of satellites carrying instruments provided by the Los Alamos National Laboratory in New Mexico. Each launch in the

Table 3.2
Satellites used to study gamma ray bursts

Satellite	Launch date	Launcher	Perigee	Apogee	Incl.	Notes
OGO-3	6 Jun 1966	Atlas Agena B	274	121 936		See *Note 1*
Explorer 34	24 May 1967	Delta	154	131 187	67.1	IMP-6. See *Note 1*
Vela 5A	23 May 1969	Titan 3C	94 052	128 529	56.4	Vela 9 ⎱ Military. Discovered GRBs. Orbits
Vela 5B	23 May 1969	Titan 3C	89 999	133 011	56.2	Vela 10 ⎰ 180° apart
Explorer 41	21 Jun 1969	TIAD	210	132 885	83.4	IMP-7. See *Note 1*
Vela 6A	8 Apr 1970	Titan 3C	106 367	116 056	54.9	Vela 11 ⎱ Military. GRB Experiment
Vela 6B	8 Apr 1970	Titan 3C	105 560	117 188	54.8	Vela 12 ⎰
OSO-7	29 Sep 1971	LTTA-Delta	201	355	33.1	See *Note 1*
Cosmos 461	2 Dec 1971	C1	303	318	69.2	Military. Decayed 21 Feb 1979
Apollo 16	16 Apr 1972	Saturn 5	Manned lunar mission			Detected GRB with lunar gamma ray spectrometer
Prognoz 2	29 Jun 1972	A2e	550	200 000	65.0	See *Note 1*. Solar wind studies plus SIGNE 1 experiment. France/USSR
Helios 2	15 Jan 1976	Titan Centaur	In solar orbit			Germany/USA solar probe. GRB detector
Gamma D2B	17 Jun 1977	C1	456	515	50.6	SIGNE 3 France/USSR
HEAO-1	12 Aug 1977	Atlas Centaur	424	444	22.7	GRB seen in Hard X-ray/Gamma Ray Expt
Prognoz 6	22 Sep 1977	A2e	488	197 867	65.0	Carried SIGNE 2MP experiment. France/USSR
Pioneer Venus Orbiter	20 May 1978	Atlas Centaur	In Venus orbit			Carried GRB Detector
ISEE-3	12 Aug 1978	Delta	Diverted to intercept comet Giacobinni–Zinner			Carried GRB detector
Venera 11	9 Sep 1978	DIE	In Venus orbit ⎱			KONUS & SIGNE 2MS GRB detectors
Venera 12	14 Sep 1978	DIE	In Venus orbit ⎰			
Prognoz 7	30 Oct 1978	A2e	483	202 465	64.9	SIGNE 2MP. Operated 7 Months. France/USSR
Solar maximum mission	14 Feb 1980	Delta	537	28.5	28.5	GRB Detector
Venera 13	30 Oct 1981	Dle	In Venus orbit			KONUS GRB Expt
Venera 14	4 Nov 1981	Dle	In Venus orbit			KONUS GRB Expt
Prognoz 9	1 Jul 1983	A2e	380	720 000	65.5	SIGNE 2MP-9 France/USSR

Note 1. Gamma ray bursts detected in data from an instrument designed for some other purpose.
Note 2. Since satellite orbits change other because of atmospheric drag etc. orbital parameter quoted by different sources may vary slightly.

GRB = Gamma ray burst
TAID = Thrust augmented improved Delta
LTTA = Long tank thrust augmented

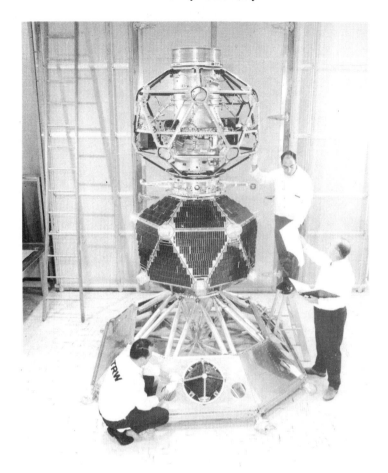

Two Vela satellites as they appeared in their launch configuration. Note the small device between the two satellites to adjust the spin rates prior to separation. (TRW).

Vela programme placed two satellites into circular orbits, 112 000 km high (see Table 3.2), positioned so that the satellites in each pair were separated by 180°. Each could view one hemisphere of the Earth together with much of near-Earth space and the area hidden from one satellite was always observable from the other. The nomenclature of the satellites is confusing; some sources refer to, for example, Vela 1 and 2; others refer to the same objects as Vela 1A and 1B. Similarly, the last two satellites, Vela 6A and 6B, are sometimes known as Vela 11 and 12.

The first three pairs of Vela (the Spanish word for vigil) satellites were 20 sided polyhedra about 1.25 m in diameter and weighing about 230 kg. They were spin stabilised at 120 rpm. The final three pairs were 26 sided with masses of about 260 kg but span at only 1 revolution per minute. The rotation axes of the last three pairs were actively pointed at the Earth with gas jets which were commanded to fire whenever the pointing error exceeded 1° (the Earth subtends an angle of only 6° from the height of a Vela satellite). In each case solar panels covered all but two faces of the satellite. The instrumentation carried varied, but included detectors sensitive to the pulse of neutrons,

X-rays, and gamma rays produced during nuclear explosions. These detectors were inactive until triggered by a flux of X- and gamma rays which exceeded a preset value. This triggering system made them sensitive to astronomical sources which produced sudden and intense bursts of high energy radiation. The Vela satellites also served as solar observatories, monitoring the X-rays and particles emitted during solar flares and providing radiation data in support of manned space missions.

The detection of X-ray transients by the Velas has already been mentioned, but the Velas also revealed a totally new phenomenon: intense bursts of celestial gamma rays, first detected in 1969 but not announced until 1973. The satellites were Velas 5A, 5B, and 6A, 6B, each of which carried six scintillation detectors with a total volume of caesium iodide (CsI) crystal of about 60 cm^3. The detectors operated over the range 150–750 keV (Vela 5A, B) and 300–1500 keV (Vela 6A, B). Once triggered they were able to monitor a gamma ray burst in some detail, recording, amongst other data, the precise time of the event. If an event was detected by more than one spacecraft, triangulation could be used to locate the direction of the source. Although the positional information provided by triangulation was limited, it was sufficient to show that the bursts originated outside the solar system.

3.4.2 Other gamma ray burst detectors

Soon other bursts were discovered, recorded by satellites like OGO-3, Explorer 34 (a geomagnetic fields spacecraft also known as Interplanetary Monitoring Platform-6), OSO-7, and even the gamma ray spectrometer on the Apollo 16 spacecraft *en route* to the Moon. However, to improve the accuracy of the position estimates it was essential to place gamma ray detectors on interplanetary spacecraft, increasing the baseline available for triangulation. A variety of detectors for this purpose have been developed; a few are described below.

The first deep space gamma ray burst detector in the NASA programme was on the Pioneer Venus Orbiter, a mission launched in May 1978 to conduct radar studies of Venus. The gamma ray instrument, restricted in size and power consumption because it was not part of the main planetary payload, used two caesium iodide scintillator detectors, 3.8 cm in diameter and 3.2 cm long, surrounded by a plastic anti-coincidence shield. The modules were installed on opposite sides of the cylindrical, spin stabilised, spacecraft. The instrument was activated 24 hours after launch and operated throughout the mission, detecting its first burst just three hours after the experiment was switched on (although nothing further was seen for several months).

A germanium crystal gamma ray detector, providing energy resolution of about 10 keV over the range 200 keV–3 MeV, was installed on the International Sun Earth Explorer (ISEE)-3 spacecraft, launched in 1978. ISEE-3 was initially placed in a halo orbit around a Lagrangian point, 230 Earth radii on the sunward side of the Earth–Sun line. This kept the spacecraft in a fixed position relative to the Earth from where it could monitor the solar wind. ISEE-3 was subsequently manoeuvered out of the Earth–Moon system and sent to intercept Comet Giacobinni–Zinner in 1985.

The KONUS system, flown on the Soviet Venera 11, 12, 13, and 14 Venus probes, consisted of a set of six scintillator detectors using sodium iodide (NaI) crystals containing traces of the rare element tantalum. The crystals were 8 cm in diameter and 3 cm thick. The detectors pointed in different directions and were shielded by lead and

The French SIGNE-3 Satellite. Note the similarity to D2B-AURA (Chapter 4). (CNES).

tin. The background counting rate was monitored regularly, and whenever a burst was detected, the last few seconds of data, plus the next 2.5 minutes' worth, was recorded for transmission back to Earth. Time resolution was highest for the first few seconds after the burst was detected, and then reduced over the remaining time in which data was recorded.

Gamma ray burst detectors were also carried by elements of the Franco-Soviet SIGNE program. SIGNE 1 was a package of experiments carried on the Earth orbiting Prognoz-2 satellite, SIGNE 2 was carried on Prognoz 6. SIGNE 3 was a small French built satellite also known as Gamma D2B (similar to the ultraviolet D2B satellite; see section 4.6.5) originally designed for the French national launcher Diamant, but transferred to a Soviet launch after the termination of the Diamant programme. Other SIGNE packages were carried on Prognoz-7, Prognoz-9, and the interplanetary Venera 11 and 12 spacecraft. Data from Prognoz-9 were combined with data from ISEE-3 to locate a gamma ray source in the direction of the constellation Sagitarius which has since been observed to burst over 100 times.

Gamma ray burst detectors were also carried on HEAO-1, the US/German Helios-2 interplanetary probe, and the Solar Maximum Mission spacecraft. Other detectors will be carried on the GRANAT and GAMMA-1 missions (see section 3.7), the Ulysses international solar polar mission, and the Soviet Phobos mission to Mars. The Phobos mission will carry two Franco-Soviet gamma ray burst experiments; LILAS operating from 3-1000 keV and APEX which covers the energy range 5 keV–10 MeV.

3.4.3 The source of the gamma ray bursts

Despite the hundreds of bursts recorded since 1969, astronomers still know very little about their origin. It seems that each burst is different, although the general characteristics can be summarised as follows. The energy spectrum of the bursts cover the region from a few tens of keV up to a few MeV, which may mean that they are an extension into the gamma ray region of the phenomena encountered in X-ray burst sources. The bursts increase to peak intensity over a period of 10–1000 milliseconds and last for periods ranging from a few tens of milliseconds to several hundred seconds. Many bursts show spectral features such as emission lines at high energies and absorption lines at low energies. The detailed interpretation of these spectra is still not fully understood.

Only one gamma ray burst source has been identified with another astronomical object; a spectacular burst which occurred on 5 March 1979 was detected by no fewer than nine spacecraft, and has been linked to a supernova remnant in the Small Magellanic Cloud. This burst did, however, have several rather unusual characteristics, and may not be typical. It is also possible that the event did not occur in the SMC, but was a chance superposition of a nearby object which lay along the line of sight. A few gamma ray bursts have been linked with X-ray sources, and there is a tentative link between gamma ray bursts and short, optical flashes detected by searching old astronomical photographs of regions known to contain a gamma ray burster. The bursters so far discovered seem to be distributed randomly across the sky, which implies that they are either very close, otherwise they would seem to cluster along the galactic plane, or very distant, since there is no evidence of any concentration in the direction of nearby clusters of galaxies such as the one in Virgo. On balance it seems that the bursts arise in the vicinity of the Sun because if they were extragalactic the energies involved would be fantastically large and very difficult to explain.

Models of the gamma ray bursters currently favour magnetised neutron stars which burst by some as yet poorly understood process. Energy generation due to internal rearrangements of the neutron star (starquakes), thermonuclear explosions of material spiralling onto a neutron star from a companion (rather like the X-ray bursters), and the energy liberated during a collision between a neutron star and a comet or asteroid have all been proposed. Evidence that the bursts might involve neutron stars comes from the fact that the signal detected from one burst showed an 8 second periodicity, consistent with the spin rate of an old neutron star. The observational details and theoretical models of the gamma ray bursters are covered in detail in *Gamma ray astronomy* by Ramana Murthy (see *Bibliography*), and will not be discussed further here.

3.5 THE SURVEYS

The pioneering experiments described in section 3.3 showed the potential for a sensitive sky survey, and two such missions took place in the 1970s. Both used similar instruments, so it is convenient to discuss the two satellites before examining the results which they obtained.

3.5.1 SAS-2

The second Small Astronomy Satellite used a spacecraft similar to the Uhuru X-ray mission but equipped with a spark chamber telescope. SAS-2 was spin stabilised and could be pointed by magnetorquers with an accuracy of about 0.3°. Data were stored on redundant, continuous loop tape recorders operating at 1 kilobit per second and

An Artist's impression of the SAS-2 satellite in orbit. (NASA).

were transmitted to a ground station at Quito, Ecuador once per orbit. Individual gamma ray detections could be timed with an uncertainty of about 1 millisecond.

The telescope consisted of a 16 module spark chamber above a set of plastic scintillators, followed by another 16 module spark chamber and a further set of detectors. The bottom detectors were Cherenkov counters, devices similar to scintillation counters, but which only produce a flash of light when charged particles, for example an electron–positron pair, travel through them faster than the speed of light in the material. The individual modules of the spark chamber contained two grids, each of 200 evenly spaced parallel wires, with the direction of the wires being different in the two grids. Each of the wires in the grid was terminated by a magnetic core, the condition of which was changed by the current which flowed if the wire was struck by a spark. By detecting which magnetic cores had been changed, the track of the spark in the module could be deduced. The position of the spark in subsequent chambers gave the paths of the electron pair in three dimensions.

The spark chamber modules were separated by thin tungsten plates which provided an environment in which pair production could occur. Subsequent scattering of the electron pair as they passed through the plates allowed the energy of the original gamma ray to be estimated. The lower chamber was included to provide additional energy information by further scattering of the electron pair. The entire detector assembly (except the bottom) was surrounded by a plastic scintillator anti-coincidence dome. The instrument logic recorded only events which failed to trigger the anti-coincidence shield but subsequently registered in both the central plastic scintillator and the lower Cherenkov counters. The angular resolution of the telescope was a few degrees, and it could detect photons with energies greater than about 25 MeV.

The satellite was launched into a low equatorial orbit from the San Marco platform on 15 November 1972, and the experiment was activated four days later. Initial observations concentrated on the region of the galactic plane, with some coverage of high galactic latitudes to search for possible point sources. Unfortunately, a failure of the input portion of the low voltage power supply terminated the mission after only seven months. By this time about 55% of the sky, including most of the galactic plane, had been observed. About 8000 photons were detected during the mission.

3.5.2 COS-B

In 1969 the European Space Research Organisation, now incorporated into ESA, conducted studies of a number of possible science missions. Amongst these was COS-A, a combined X-ray and gamma ray mission, and COS-B, a gamma ray satellite carrying a single experiment. COS-B was approved for development, and an international consortium of five research groups, calling themselves the Caravane Collaboration, produced the scientific payload. The British group was forced to withdraw owing to lack of funds and was replaced by the Space Science division of ESTEC, an ESA institution located in the Netherlands.

The spacecraft was designed to be simple and to be compatible with both the ill fated Europa II European launcher and the American Thor–Delta rocket. The result was a cylindrical spacecraft, spin stabilised at about 10 rpm, with solar cells around the circumference of the cylinder. The gamma ray telescope, a spark chamber, was fitted in the centre of the spacecraft with its field of view centred on the spin axis. A small X-ray detector, also looking along the spin axis, was included in the hope that simultaneous X- and gamma-rays could be detected from pulsars. The satellite was pointed by a simple nitrogen gas jet system, with Earth and Sun sensors used to determine the pointing direction. No data storage system was carried, since COS-B was placed in a highly eccentric (apogee 100 000 km) polar orbit which kept it in direct contact with ground controllers much of the time. The total mass was 278 kg, of which 118 kg was the experiment.

The COS-B telescope was broadly similar to the SAS-2 experiment. It consisted of a 16 grid spark chamber, with tungsten sheets interleaved between the top 13 grids, mounted above a set of scintillation and Cherenkov trigger detectors. The spark chamber was filled with neon gas (plus a little ethane) at a pressure of 2 bar. A gas flushing system was provided to empty the spark chamber and refill it with fresh gas to prevent deterioration of the experiment due to contamination of the chamber gas. As in SAS-2, the path of a spark through the chamber was detected by magnetic cores threaded on the ends of the wires. The spark chamber was surrounded by a plastic scintillator anti-coincidence dome monitored by a set of nine photomultipliers. Correct operation of the anti-coincidence system was essential since the highly eccentric orbit of COS-B took the satellite well beyond the Earth's radiation belts, exposing the instrument to a large background flux of charged particles and cosmic rays.

Below the spark chamber and its triggering detectors was an energy calorimeter, a caesium iodide crystal scintillator capable of totally absorbing electrons with energies of up to 300 MeV. Electrons able to penetrate the crystal and escape were detected by another plastic scintillator. This final detector allowed the experimenters to record the number of very high energy events and to place a lower limit on their energy.

COS-B was launched on 9 August 1975 and had a design life of two years. The satellite was operated in a pointed mode, with the viewing direction fixed on a single direction for periods of between 30 and 75 days. On some occasions the pointing direction was such that the small X-ray detector was able to monitor known X-ray sources, and X-ray data were obtained for a small number of sources. COS-B proved more durable than SAS-2, and the spark chamber gas remained cleaner than expected, so the satellite continued to operate until 25 April 1982 when the chamber gas supply finally ran out and the instrument was deactivated. COS-B detected about 100 000 gamma ray photons during its 6.5 year operating life.

3.5.3 The gamma ray sky[†]

Both SAS-2 and COS-B produced maps of gamma ray emission. The SAS-2 experiment confirmed than the galactic plane was a source of gamma rays, and showed that the emission extended considerably on either side of the galactic centre. The COS-B satellite, because of its much longer operating life, was able to extend the SAS-2 observations, and the Caravane Collaboration has published a number of detailed maps of diffuse gamma ray emission covering several different energy bands. These show an intense ridge extending about 60° either side of the galactic centre. Since the ridge is narrow it is believed that much of the emission arises from cosmic ray interactions with interstellar gas clouds between the Sun and the centre of the Galaxy. Additional gamma rays are produced by radiation processes, for example synchrotron radiation, involving the electron component of the cosmic rays. Although there appears to be some emission from higher galactic latitudes, the question of whether it is diffuse extragalactic emission or is due to distant point sources, remains controversial.

The poor angular resolution of both SAS-2 and COS-B makes the search for point sources of gamma ray emission very difficult. The SAS-2 experimenters successfully identified gamma rays from two pulsars, the Crab Nebula and the Vela pulsar (named after the constellation Vela, not the Vela satellites), detections confirmed by the discovery that the gamma rays were pulsed with the same periods as the pulses reported by radio astronomers. The SAS-2 group also suggested that they had detected two other pulsars, but these identifications have not stood the test of time.

The SAS-2 data hinted that several other point sources might exist, including a strong unidentified source in the constellation of Gemini, and possible emission from the X-ray source Cygnus X-3, but definite identifications were not possible. The much larger database provided by COS-B allowed the identification of about 25 pointlike sources including the Crab and Vela pulsars and the SAS-2 source in Gemini, informally christened 'Geminga' (which in Milanese dialect means 'non-existent') by Giovanni Bignami, one of the scientists of the SAS-2 mission. The detection of Cygnus X-3 was not confirmed, but one source has been identified as the nearby quasar 3C 273.

Geminga is most unusual. In the SAS-2/COS-B energy range it is the second most

---

[†]For a detailed review of gamma ray astronomy see *Gamma-ray astronomy* by P. V. Ramana Murthy and A. W. Wolfendale.

powerful gamma ray source known, but despite many hours of COS-B observations it proved impossible to link it to any other object. A sensitive search of the COS-B error box with the Einstein observatory revealed the existence of an X-ray source with no obvious optical counterpart, but powerful optical telescopes finally revealed a faint, magnitude 25, star. X-ray observations with EXOSAT, which showed evidence of variability with a period of about 1 minute, roughly in agreement with estimates of the gamma ray variability, seemed to confirm the identification. X-ray and optical measurements indicate that Geminga is close, probably within 200 parsecs of the Sun.

Some astronomers have suggested that the period of Geminga is slowly changing, having increased from just over 59 seconds in 1972 to rather over 60 seconds by 1983. If this is true (and not everyone agrees that it is) it will hinder the comparison of data taken at different times, and it has heightened the mystery surrounding Geminga. The nature of the object is unclear, but current models favour a binary system containing two magnetised neutron stars even though the likelihood of finding such a system so close to the Sun is very small indeed. Much more work is required to confirm if this idea is correct.

It remains to be seen if further analysis will lead to any more identifications of COS-B sources. The problem is compounded by the very poor angular resolution of the telescope, which means that some of the COS-B sources may not be discrete objects at all, but may be due to cosmic ray interactions with clumps of interstellar gas. One COS-B source has already been linked to the cloud of gas and dust near the star Rho Ophiucus. Presumably cosmic ray interactions with dense regions within the cloud produce gamma ray emission which mimics that expected from a point source. Until a new generation of gamma ray telescopes capable of much greater angular resolution are launched it seems likely that many of the sources in the COS-B catalogue will remain unidentified.

3.6 HEAO-3 AND GAMMA RAY SPECTROSCOPY

The production of gamma ray spectral lines was mentioned in section 3.1, but gamma ray spectroscopy is hampered by the very low fluxes reaching the Earth and by the ever present gamma ray background. Although most gamma ray spectroscopy has been done from balloons, a number of results have been obtained by small experiments on satellites such as Cosmos 856, 914, and 1106 (see Table 3.1) and by the spectrometer on HEAO-3, the third High Energy Astrophysical Observatory.

The first two HEAO satellites (see section 2.4.6) were X-ray astronomy missions, but the final HEAO carried experiments devoted to higher energies. HEAO-3 used the same spacecraft as its predecessors, but carried a gamma ray spectrometer and two cosmic ray instruments. HEAO-3 was launched on 20 September 1979 into a near circular low Earth orbit where it rotated slowly around its Sun facing axis, keeping its solar panels illuminated while scanning its scientific instruments across the celestial sphere.

The HEAO-3 spectrometer covered the range 0.6–10 MeV and used four solid state germanium detectors, cooled to about 95 K by a refrigerator containing reservoirs of frozen ammonia and frozen methane. The amount of coolant carried was based on the planned six month survey mission, but actually lasted rather longer. A set of CsI (Na) scintillators were fitted around the instrument to act as both an anti-coincidence shield and as an active collimator with a total field of view of about 30°. The instrument initially worked well, but cosmic ray interactions with the germanium detectors gradually degraded the energy resolution from about 3.5 keV just after launch to 19 keV about six months later. Despite this, a number of significant measurements were made.

The 511 keV line produced during electron–positron annihilation was observed from the galactic centre during the later months of 1979 and the earlier months of 1980. The flux appeared to be variable over periods of a few months, implying that the emitting region is probably less than 0.3 parsecs across. A possible explanation of this is the presence of a black hole near the galactic centre; hot gas surrounding the black hole might emit Bremsstrahlung photons which could then collide to produce electron–positron pairs which would in turn annihilate to produce the 511 keV spectral line.

HEAO-3 experimenters also found a line at 1.809 MeV, which arises when an unstable isotope of aluminium (^{26}Al) decays into magnesium. Aluminium 26 has a half life of about one million years and is present in the interstellar medium only because new supplies are regularly produced during supernovae. The HEAO-3 observations seem to suggest that there is more interstellar ^{26}Al than expected, which may mean that there is another, yet to be discovered, source of this isotope. Unfortunately the signal from the ^{26}Al was so weak that the spatial distribution of the emission could not be mapped, and no further clues to its origin are available from the HEAO-3 data.

Gamma ray lines at 1.2 and 1.5 MeV were seen from the bizarre object SS433, thought to be a binary system containing a black hole accreting material from a companion star and at the same time ejecting powerful jets of material at right angles to its accretion disk. The lines are believed to be from an isotope of magnesium (^{24}Mg) which can produce a 1.369 MeV line by the relaxation of an excited nuclear state. In SS433 the line is both red and blue shifted from its normal energy owing to the very high velocity of the jets (for more details of this fascinating object see *The quest for SS433* by David Clarke). No spectral lines were detected from other discrete sources, although searches were made for lines arising in objects such as the Crab Nebula and unusual galaxies.

3.7 FUTURE GAMMA RAY MISSIONS

3.7.1 GAMMA-1

GAMMA-1 was a Soviet mission conceived in late 1972 to investigate the energy range from 50–5000 MeV but which was widened in 1974 to include three French research centres. Poland is also taking part in the mission, being responsible for the TELEZVEZDA star sensor.

The main experiment is a 12 gap spark chamber, 0.5 m square, mounted above a pair of plastic scintillators and a gas Cerenkov counter which together act as trigger detectors. Below them is a two gap spark chamber (to help track the electron–positron pair after they leave the main spark chamber) and an energy calorimeter, constructed from a stack of 24 scintillation counters and 24 sheets of lead, monitored by photocells. A vidicon (TV) system is used to measure the positions of sparks in the main chamber to an accuracy of 0.125 mm, and, as usual, the detector assembly is surrounded by an anti-coincidence shield. The angular resolution of the instrument is expected to be better than 2° for photons with an energy of 100 MeV. Improved angular resolution can be obtained by positioning a coded aperture system, consisting of two one-dimensional arrays made from 1 cm thick tungsten, in front of the detector. The mask, which was added late in the development programme, is rather close to the spark chamber, and this limits the improvement in angular resolution to about 20 arc minutes, rather less than the resolution possible if the mask to detector spacing was somewhat larger.

In addition to the spark chamber two smaller instruments are included in the payload. The PULSAR X-2 instrument is optimised for observations of X-ray pulsars and consists of a set of four proportional counters covering the 2–25 keV energy range. The instrument should be able to study X-ray pulsars with periods ranging from a few microseconds to a few days. The second instrument is a small, mechanically modulated, crystal scintillator called DISK designed to detect weak gamma ray sources of intermediate energy (20 keV–5 MeV) against the background of more intense emission from both diffuse gamma ray emission and charged particles.

The GAMMA-1 payload will be carried aboard a modified Soyuz spacecraft, into a 400 km high circular Earth orbit inclined at about 51° to the Equator. Launch should occur in 1988. The satellite will be three axis stabilised with an accuracy of about 0.5°. Final data processing, using data from the TELEZVEZDA star sensor which has a 6° square field of view and can record 7th magnitude stars, will be used to determine the spacecraft pointing direction with an accuracy of about 2 arc minutes. The satellite will operate by making uninterrupted observations in a single direction for periods of up to one month. Data will be transmitted to the ground twice a day.

3.7.2 GRANAT

GRANAT is a Soviet Gamma ray mission with considerable international collaboration. The largest experiment is a French built coded mask telescope called SIGMA, but Denmark and Bulgaria are also concerned with other aspects of the project.

The SIGMA telescope is designed for high resolution imaging in the energy range 30 keV–2 MeV and was conceived by a European group including France, the UK, and Italy who proposed it as a payload for the first launch of the Ariane 4 rocket, a fact reflected in the experiment's original designation—Satellite d'Imagerie Gamma Monté sur Ariane. However, the French group withdrew from the consortium and entered into an agreement with the Soviet Intercosmos organisation under which the telescope was incorporated into the GRANAT mission. SIGMA (redesignated Systeme d'Imagerie Gamma à Masque Aléatoire) will be used to study targets such as the galactic centre, the Rho Ophiucus gas cloud complex, and various active galaxies and quasars.

The 900 kg SIGMA telescope uses a coded mask technique similar in principle to the one carried on Spacelab 2 (see section 2.5.4). A tungsten mask, 1.5 cm thick and approximately 30 cm square, is supported 2.5 m from a position sensitive detector. Photons cast a shadow of the mask onto the detector, and the resulting pattern is deconvolved by computer. The detector is a 1.25 cm thick NaI (Tl) crystal scintillator viewed by 61 photomultipliers, and it is surrounded on all but one side by a CsI anti-coincidence detector. The mask, coded with a repeating pattern of 29 × 31 elements in a 27.8 × 29.7 cm rectangle, is held in position by an aluminium tube. The angular resolution of the telescope is expected to be about 1 minute of arc.

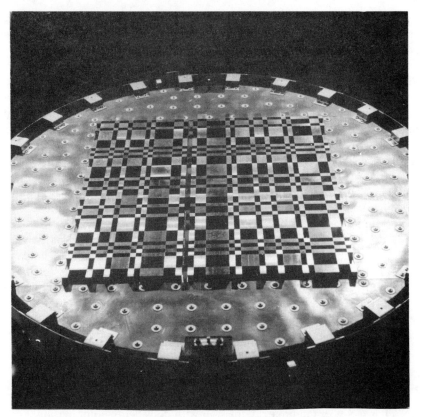

A close up of the mask of the SIGMA telescope. (CNES).

To obtain the maximum possible sensitivity the SIGMA telescope needs to make long pointed observations from a vantage point at which the properties of the background radiation remain stable. To meet these criteria, GRANAT, which is similar to the ASTRON spacecraft described in section 4.8, will be launched into a highly elliptical orbit (2000 × 20 000 km, inclination 51°) with a period of about 4.5 days. During each orbit the satellite will spend about 4 days above 6000 km altitude where the background radiation will be fairly constant. This orbit will allow near continuous visibility to Soviet ground stations, but observatory operations are not planned; instead, the telescope will make pre-programmed observations of targets selected by a Franco–Soviet science team.

GRANAT will also carry a suite of other instruments. ART-P is a hard X-ray (4–100 keV) experiment which uses four small coded mask telescopes operating in the 3–100 keV energy range. Each of the co-aligned telescopes has a field of view of about 2° square and can produce an image with an angular resolution of 5 arc minutes. ART-S is able to carry out moderate resolution X-ray spectroscopy in the 3–150 keV energy range and to monitor X-ray bursts.

The other experiments are X-ray/Gamma ray burst detectors. The KONUS experiment is similar to those flown on the Venera missions and operates from 20–800 keV. The KONUS package is linked to a device known as PODSOLNUKH which consists of two proportional counters mounted on a movable platform. When the KONUS system detects an X-ray burst the PODSOLNUKH platform will turn in a matter of seconds to study the source in the 2–25 keV range. A French gamma ray burst instrument called PHOEBUS will operate from 100 keV–100 MeV, and a Danish all sky monitor called WATCH will cover the range 5–120 keV. The WATCH instrument consists of four modules and uses rotation modulation techniques to assist in the location of sources with an accuracy of less than 0.1°.

3.7.3 The Gamma Ray Observatory (GRO)

The GRO is a NASA facility class mission planned for launch after 1990. The 17 tonne GRO will be placed by the Space Shuttle into a parking orbit from which it will use its own hydrazine propulsion system to climb to a 400 km orbit inclined at 28.5° to the equator. At least 4 kW of electrical power will be provided by two solar arrays which stretch about 23 m from tip to tip and can be restowed or jettisoned as required for servicing. Communications will be via the TDRSS relay system. Like the HEAO satellites, the GRO will be developed in accordance with the protoflight philosophy; there will be no engineering or qualification models, just a single example built to fly. The observatory will operate in a pointed mode, probably spending two weeks observing in a single direction before moving to a new observing attitude. The pointing directions will be selected so that the two instruments with wide fields of view

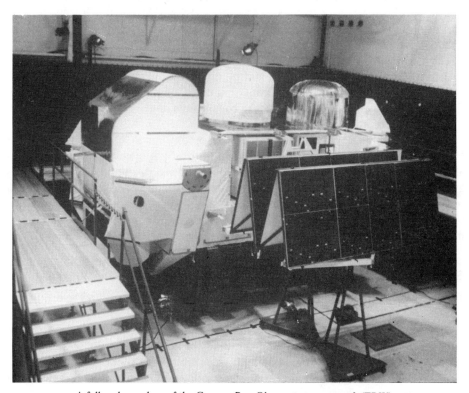

A full-scale mockup of the Gamma Ray Observatory spacecraft (TRW).

(COMTEL and EGRET) can observe the entire sky during the mission. The initial mission will last for about two years before the GRO is either serviced in orbit, returned to Earth for refurbishment, or allowed to burn up during re-entry.

The objectives of the GRO include a survey of diffuse and pointlike gamma ray sources, the detection and identification of gamma ray spectral lines, and studies of gamma ray bursts. To achieve these the satellite will carry four experiments, selected in 1981, which are summarised below. The energy range of each instrument is chosen to overlap with at least one of the others to ensure complete spectral coverage.

The Burst Transient Source Experiment (BATSE) is designed to detect gamma ray bursts in the energy range 0.05–1 MeV. It consists of eight detectors—four viewing parallel to the other instruments, four observing the opposite hemisphere. The BASTE is expected to locate burst sources with an accuracy of 1° and has a timing accuracy of about 100 microseconds.

The Compton Telescope (COMPTEL) detects gamma rays by observing the electrons produced during Compton scattering of low energy (1–30 MeV) photons in the target material. This GRO instrument, which is provided by a European consortium, uses two detector modules 1.5 m apart, an upper liquid scintillator, and a lower NaI(Tl) crystal, to provide improved angular resolution. Even with this refinement the rather ill defined relationship between the path of an incident photon and that of each scattered electron means that the COMPTEL will have a resolution of only a few degrees.

The Energetic Gamma Ray Telescope (EGRET) is a multilayer spark chamber plus energy calorimeter similar to SAS-2, COS-B, and GAMMA-1, but larger and with improved angular resolution—about 10 arc minutes for a few strong point sources and about 1° for most other objects. It operates in the 20–30 000 MeV range.

The Oriented Scintillation Spectrometer Experiment (OSSE) uses four actively shielded and passively collimated NaI scintillators operating between 0.05 and 10 MeV. The detectors have independent, single axis, orientation systems which allow them to look away from a source and take background data if required. The OSSE has a field of view of about $5 \times 11°$.

GRASP, Gamma Ray Astronomy with Spectroscopy and Positioning, is a proposed European mission using a sensitive coded mask telescope operating between 15 keV and 100 MeV. The use of a coded mask telescope, field of view about 50 square degrees, and a detector with high spectroscopic resolution, would make this instrument capable of taking high quality spectra of many individual sources simultaneously. If developed, GRASP could provide, in only four months, a survey of the galactic plane an order of magnitude better than that produced during the 6.5 year COS-B mission.

Some highlights of space gamma ray astronomy

1958 Astronomer Philip Morrison predicts several mechanisms for the production of gamma rays by astronomical sources.
1961 Explorer 11 carries small scintillation counter into orbit.
1967 OSO-3 detects gamma rays from the galactic plane.
1972 SAS-2 gamma ray survey. Satellite fails after seven months.
1973 Discovery (in 1969) of gamma ray bursts by military Vela satellites is announced.

1975 European COS-B satellite launched on sky survey mission; COS-B operates successfully for six years.
1979 HEAO-3 launched with a payload of a gamma ray spectrometer and two large cosmic ray experiments.
Single gamma ray burst, possibly occurring in the Large Magellanic Cloud, is detected by nine different satellites.

4

Ultraviolet and extreme ultraviolet astronomy

4.1 INTRODUCTION

Just as hard X-rays merge into gamma rays, progressively softer X-rays enter a no-mans-land which links them to the ultraviolet. Portions of this transition region are known, depending on the astronomer concerned and in order of decreasing photon energy (increasing wavelength), as the X-ray Ultraviolet (XUV), the Extreme Ultraviolet (EUV), and the Far Ultraviolet (FUV). The term Vacuum Ultraviolet (VUV) is also sometimes used. Astronomy at these intermediate wavelengths is hampered by both technical and physical obstacles, so they are areas where detailed observations are only just beginning. At longer wavelengths lies the ordinary ultraviolet region with which the major part of this chapter is concerned and which is bounded by two fairly distinct limits.

The short wavelength limit of the ultraviolet can be regarded as about 91 nm. At wavelengths shorter than this, individual photons are energetic enough to eject an electron from a normal hydrogen atom and ionise it. Since interstellar space contains large quantities of neutral hydrogen, each atom of which can soak up a suitable photon, the visibility at shorter wavelengths (that is, in the FUV/EUV/XUV region) is severely limited by a fog of neutral hydrogen. Only when the wavelength becomes so small that the probability of a collision between a photon and a hydrogen atom falls dramatically does the range of vision increase.

The long wavelength limit is set by the ozone in the upper atmosphere which absorbs photons with wavelengths shorter than about 310 nm. This marks the shortest ultraviolet wavelengths which can penetrate to ground level and be studied without recourse to rockets or satellites. Strictly speaking the human eye perceives the violet end of the spectrum as about 390 nm, not 310 nm, but since ground based electronic detectors can work between 310 and 390 nm it seems sensible to regard this region as the domain of terrestrial telescopes.

Astronomers are interested in the ultraviolet because this region contains many important spectral lines, notably from heavy elements such as carbon, nitrogen, and oxygen. Observations of these lines, impossible from the ground, can provide information about the chemical composition of all kinds of astronomical objects including comets, planets, stars, galaxies, and the interstellar medium. In hot environments, such as stars and the expanding clouds of gas surrounding supernovae, the spectral lines appear in emission, as electrons in atoms are excited to higher energy states and then fall back, emitting an ultraviolet photon in the process. Spectral lines are also seen in absorption when ultraviolet light passes through material in interstellar space. In this case, when the light from a distant star encounters the interstellar material, photons of suitable energies are absorbed and this removes photons with characteristic wavelengths and produces dark lines in the spectrum of the star. Once due allowances have been made for absorption features in the original stellar spectrum, the composition of the absorbing material can be fingerprinted with considerable precision.

Ultraviolet measurements are also important for the study of very hot stars. Stars with masses ten or more times that of the Sun must burn their nuclear fuel very quickly to support themselves against the force of gravity, and so have surface temperatures in the region of 10 000 to 100 000 degrees. These short lived stars emit much of their light in the ultraviolet, so astronomers must concentrate their observations of these objects at ultraviolet wavelengths.

The existence of stars likely to have considerable emission in the ultraviolet, and the presence of important ultraviolet spectral lines from heavy elements, was of course well established by optical astronomy and laboratory studies. It was thus clear, even before a single observation had been made, that ultraviolet astronomy was full of possibilities. Inevitably, it was one of the first wavebands to be exploited from space.

4.2 ROCKET OBSERVATIONS

Since the height of the ozone layer was not known until the advent of sounding rockets, there were a number of attempts to obtain ultraviolet spectra of the Sun from manned balloons. Most of these flights took place in the 1920s, but even as late as the 1930s it was still believed, incorrectly, that balloons might rise high enough to penetrate the ozone layer. The true situation was finally revealed by the V2 rockets launched from White Sands in 1946, and it was from one such rocket that Richard Tousey and his group managed to obtain a photograph of the solar spectrum which stretched well into the ultraviolet.

Over the next two decades rocket flights observed the Sun at steadily shorter wavelengths, but it was not until the late 1950s that ultraviolet radiation was detected from stars. The first crude detections were made by small telescopes carried in spinning rockets, but the development of sophisticated rockets able to point at selected objects made it possible to obtain detailed ultraviolet spectra of other stars. The first to achieve this were D. Morton and L. Spitzer of Princeton University who, despite the failure of their rocket's parachute system, managed to obtain the spectra of Delta and Pi Scorpio from film recovered from their mangled spectrograph. Other successes followed, and soon several hundred stars as faint as magnitude 6.5 had been observed from rockets, leading inevitably to a strong interest in satellites devoted to extra-solar ultraviolet astronomy.

4.3 EARLY ULTRAVIOLET OBSERVATIONS FROM SATELLITES

The first satellites with experiments in ultraviolet astronomy were launched in 1964. The unnamed American satellite designated 1964-83C was launched into a near circular polar orbit 1100 km high on 12 December and carried two telescope photometers (a photometer is a device which measures the brightness of a celestial object over a limited range of wavelengths), one of which operated at a wavelength around 137.6 nm. Although the satellite could not be pointed with any accuracy, data from sensors on the spacecraft allowed the pointing directions to be established to an uncertainty of about 1° during subsequent data analysis. A list of ultraviolet observations of 96 stars made by the satellite was produced in 1966.

The Soviet Cosmos 51 satellite, launched on 10 December 1964, is reported to have observed the stellar background in the ultraviolet, and Cosmos 215, launched on 19 April 1968, was a solar observatory also equipped with a stellar ultraviolet photometer. Both missions had short lifetimes, only 72 days in the case of Cosmos 215, and little has been reported about the results of their experiments. Cosmos 262, launched on 26 December 1968, was equipped with three 16 channel photometers covering the ultraviolet and soft X-ray bands. The satellite operated for about four months and re-entered in July 1969.

4.4 THE ORBITING ASTRONOMICAL OBSERVATORIES

More significant were the American Orbiting Astronomical Observatories (OAO), a project conceived before the first satellite was even launched. In the early 1950s, when plans were being made for the American Vanguard satellite project, a number of astronomers were invited to suggest experiments suitable for small satellites. In response Drs A. Code, Goldberg, L. Spitzer, and F. Whipple proposed a number of experiments and were invited to discuss their ideas with NASA programme managers in 1959. Since it was plain that astronomical satellites would require a pointing system much more advanced than any then available, it was decided to develop a single type of spacecraft able to accommodate each of the proposed instruments in turn. As design studies of these Orbiting Astronomical Observatories progressed it became clear that Dr Goldberg's solar experiments had very different requirements, and a separate series of satellites, the OSOs (see section 2.4.1) were built for solar studies.

Despite their early start, the gestation period of the OAOs was a long one. The satellites were originally planned for a rocket known as the Vega, but this was cancelled and the design was modified to be compatible with the Atlas launcher instead. The development of star trackers and attitude control systems able to stabilise the OAOs with an accuracy of 0.1 arc second proved far more demanding than anticipated, and required a combination of momentum wheels, gas jets, and magnetorquers together with a variety of complex electronics to control them all. Eventually the OAO spacecraft evolved into an octagonal structure, about 3 m in length and 2 m in diameter with deployable solar panels mounted on opposite sides. The spacecraft subsystems were grouped around the inner circumference of the body, leaving a cylindrical space 1.25 m in diameter running down the centre of the satellite which contained the experiments assigned to each particular mission. The total mass of each OAO was about 2000 kg, and the first one, known before launch as OAO-A, was ready by early 1966.

4.4.1 OAO-1

The first OAO carried four experiments. The main payload was a University of Wisconsin package to determine the energy distribution of stars over a number of different ultraviolet wavelengths and to measure emission lines from diffuse nebulae. The instrument used a number of small telescopes equipped with electronic detectors and filter wheels. Two X-ray experiments were also fitted, a soft (2–150 keV) X-ray experiment from the Goddard Spaceflight Center and a second soft X-ray instrument from the Lockheed Corporation Research Laboratory. The fourth instrument was a 50–100 MeV gamma ray experiment provided by the Massachusetts Institute of Technology and based on the instrument flown on Explorer 11 (see section 3.3).

The satellite was launched on 8 April 1966 into a near circular orbit at an altitude of 800 km. Unfortunately, the day after launch, problems developed with the electrical systems, and overheating of the primary battery led to complications which disabled the power supply and left OAO-1 inoperable. The astronomical experiments never returned any data, and the mission was written off after three days, its only legacy a few engineering measurements which proved that the concept of an orbiting observatory was feasible.

4.4.2 OAO-2

After the failure of OAO-1, NASA implemented various modifications before launching another OAO. Originally the second launch would have been the OAO-B spacecraft, but production delays meant that the next satellite ready for launch was

The ill-fated OAO-1 satellite which failed shortly after reaching orbit. (NASA).

98   Ultraviolet and extreme ultraviolet astronomy

actually OAO-A2, the prototype spacecraft refurbished after the end of the development programme. OAO-A2, which was named OAO-2 after launch, carried two experiments, one looking out of each end of the cylindrical payload compartment.

At one end was an improved version of the University of Wisconsin instrument carried aboard OAO-1. This comprised a total of seven different telescopes grouped into three sub-packages. The main element was a stellar photometer which used four 20 cm reflecting telescopes, each equipped with a photoelectric photometer, which together covered the range from 100 nm–400 nm in slightly overlapping wavelength bands. Each telescope was equipped with filters able to further subdivide their operating region into 25 nm wide bands. A second part of the instrument consisted of two spectrometers, one on either side of the photometer package. Each spectrometer used a diffraction grating to split up the incoming light into a spectrum, part of which was then focused onto phototube detectors. By rotating the diffraction grating in steps, the spectrum could be scanned across the detectors and recorded. One spectrometer covered the wavelength range 100–200 nm, the other the 200–400 nm region. The resolutions of the spectrometers were 1 and 2 nm respectively. The third element of the package operated in the 150–380 nm band and was known as the nebular photometer. This used a 40 cm telescope and a photometer to measure the brightness of diffuse sources in five wavelength bands, each about 60 nm wide.

The other instrument was the Smithsonian Observatory Celescope, (CELEstial teleSCOPE) which was designed to survey the sky between 120 and 290 nm. Unlike the Wisconsin package, which measured only one star at a time, the Celescope imaged small areas of the sky, observing all the stars in the field of view at once. The instrument used four 32 cm telescopes each with a 2.8° diameter field of view, and each equipped with a different filter. Each telescope fed a complicated detector known as a Uvicon, which was in effect an ultraviolet sensitive television camera, able to record stars down to about 8th magnitude during a 1 minute exposure. This experiment had been

OAO-2 is prepared for launch. (NASA).

scheduled for OAO-1, but delays in development of the Uvicon caused it to be transferred to a later mission.

OAO-2 was launched on 7 December 1968 and began operations a few days later. Although one of its Uvicons was blinded by overexposure to sunlight, the Celescope obtained over 7400 images and provided useful data on over 5000 stars. The sensitivity of the Uvicons did, however, reduce as the mission progressed, and because of this Celescope observations were discontinued in April 1970, four months longer than the one year design life. The Wisconsin package remained operational for a number of years, and OAO-2 was finally deactivated in February 1973 after an electrical problem.

Apart from its successful exploratory mission of surveying the ultraviolet sky, OAO-2 was also able to observe a number of other interesting targets. These included the detection of a huge cloud of hydrogen around comet Tago-Sato-Kisaka in January 1970 and observations of a nova in the constellation of Serpens in February 1970. OAO-2 also observed globular clusters and galaxies, finding in particular that many extragalactic objects were brighter than expected in the ultraviolet. The spacecraft was even used to observe the Earth's atmosphere; by pointing at a star close to the horizon and observing it as it set behind the Earth, OAO-2 was able to determine the amount of oxygen and ozone in the upper atmosphere.

4.4.3 OAO-B

The third OAO, known as OAO-B before launch, carried a single instrument: a 1 m reflecting telescope equipped with a grating spectrograph. The main objective was to

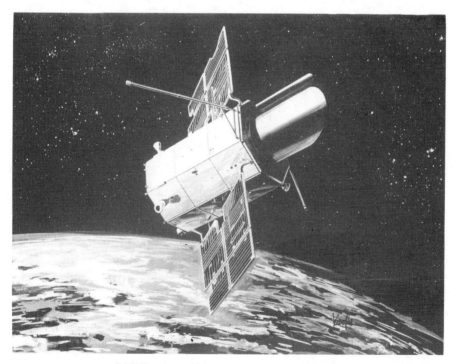

OAO-B as it would have appeared in orbit (NASA).

examine about 14 000 stars per year at high spectral resolution in order to determine the chemical and physical conditions under which various spectral lines arose. Other objectives were to study variability in classes of unusual stars, to determine the interstellar reddening law (the way in which dust clouds preferentially block blue light, making stars appear redder than they really are), and to measure the spectra of bright galaxies. OAO-B was launched on 30 November 1970 by an Atlas–Centaur rocket. Unfortunately the aerodynamic shroud designed to protect the satellite during the early portion of the launch failed to jettison, and the additional mass prevented orbital velocity being reached. OAO-B crashed into the Atlantic a few minutes after launch; it was a total loss.

4.4.4 OAO-3 Copernicus

The fourth and final satellite (OAO-C) carried an ultraviolet telescope provided by Lyman Spitzer's group from Princeton University. The instrument was an 80 cm Cassegrain telescope equipped with a high dispersion spectrograph. The spectrograph

OAO-3 (Copernicus) before launch (NASA).

used a diffraction grating ruled with 2400 lines per millimetre to disperse the incoming beam into a spectrum, and two moving detectors to scan along the dispersed beam to record the intensity at each wavelength. Each detector was sensitive to a different region, and between them they covered the range 75–300 nm at a spectral resolution of about 0.01 nm. To obtain maximum resolution the experiment carried its own fine error sensor which provided commands to the satellite's attitude control system and allowed a pointing accuracy of 0.1 arc seconds to be achieved. The satellite also carried the small, but highly successful, X-ray instrument described in section 2.4.3.

OAO-C was launched on 21 August 1972, a few months after the European TD-1A satellite (see section 4.5), and became OAO-3 on reaching orbit. The satellite was named Copernicus to commemorate the 500th anniversary of the birth of this famous astronomer, and the designation OAO-3 is seldom used. At 2220 kg Copernicus was heavier than previous OAO satellites (it was the heaviest NASA payload to that date), and as a result was placed in a slightly lower orbit than OAO-2. Checkout of the satellite took about four days at the end of which it was declared operational.

Copernicus remained in service for nine years and was used for studies of the interstellar medium, observing the absorption lines produced as starlight traversed interstellar clouds, and determining the relative abundances of the different elements in the gas between the stars. When these results were compared with the ratio of elements found in the Sun it was found that many of the elements in the interstellar gas are underabundant compared to the amount of hydrogen observed. Since the ratios of elements observed in newly formed stars (which are formed from interstellar material) are similar to the ratios found for the Sun, the elements which appear to be missing from

An impression of OAO-3 (Copernicus) in orbit.

Table 4.1.
Satellites with ultraviolet telescopes

Name	Launch Date	Launch Vehicle	Perigee (km)	Orbit[†] Apogee (km)	Inclination (°)	Notes
Cosmos 51	10 Dec 1964	B1	164	344	48.8	Decayed 14 Nov 1965. UV sky brightness experiment
1964-83C	12 Dec 1964	Thor–Able Star	1014	1073	89.8	Military, carried UV photometer
OAO-1	8 Apr 1966	Atlas-Agena D	790	800	35.0	Failed in orbit. No data returned
Cosmos 215	19 Apr 1968	B-1	162	265	48.5	UV photometer, X-ray and visible experiments
OAO-2	7 Dec 1968	Atlas-Centaur	761	770	35.0	UV observatory
Cosmos 262	26 Dec 1968	B-1	163	508	48.5	Military. Re-entry 18 July 1969. UV and X-ray experiments
OAO-B	30 Dec 1970	Atlas-Centaur		Launch failure		Aerodynamic shroud failed to jettison
Orion 1	19 Apr 1971	D1	124	130	51.5	Carried in Salyut 1
TD-1A	12 Mar 1972	Delta	531	539	97.5	European mission. UV sky survey
OAO-3	21 Aug 1972	Atlas-Centaur	729	739	35.0	Copernicus. UV observatory
Apollo 16	16 Apr 1972	Saturn 5	Manned Lunar Mission			Deployed UV camera on lunar surface
Skylab	14 May 1973	Saturn 5	422	442	50.0	Space station. 3 UV experiments
Orion 2	18 Dec 1973	A2	222	254	51.6	Carried in Soyuz 13
ANS	30 Aug 1974	Scout	256	1098	98.0	1st Netherlands satellite, UV observatory
D2B-AURA	27 Dec 1975	Diamant B	411	499	37.2	French UV mission
Prognoz-6	22 Sep 1977	A2e	488	197 867	65.0	Galactika UV spectrophotometer
IUE	26 Jan 1978	Delta	26 221	45 336	28.4	UV observatory
Astron 1	23 Mar 1983	D1e	2000	200 000	51.5	UV observatory
Spacelab 1	28 Nov 1983	Space Shuttle (Columbia STS-9)	242	254	57.0	2 UV experiments, both partial successes
Kvant	31 Mar 1987	D-1	344	363	51.6	Glazar UV telescope in Module docked to Mir space station

[†] Since satellite orbits change because of atmospheric drag etc., orbital parameters quoted by different sources may vary

the interstellar gas must be bound up in grains of dust. Because material in grains cannot produce spectral lines it is undetectable in the ultraviolet, and this leads to the apparent underabundances. The relative proportions of the 'missing' elements can then be used to estimate the composition of interstellar dust. Copernicus also enabled astronomers to probe the density of the interstellar hydrogen and to confirm that the interstellar gas is not a uniform mixture, but contains dense clouds with relatively clear voids in between. These voids are not completely empty, but contain a thin, highly ionised, gas with a temperature of about 200 000 K. This hot gas probably originated in ancient supernovae and now permeates much of our Galaxy.

Stellar observations by Copernicus also produced interesting results, notably in enabling astronomers to study the hot, outer coronae of stars. Chromospheric emission was observed from a number of cool stars, that is those of spectral types F, G, K, and M, providing new information on the outer atmospheres of these stars. Copernicus was also able to study the powerful stellar winds of hot stars, for example those of type O, and the more or less stationary gas shells around less energetic stars.

Despite the launch of various other satellites during its lifetime, Copernicus remained in service for many years because of its wide spectral coverage and very high resolution. The main limitation of the satellite was that its detectors were not suited for observations of faint stars, and the advent of satellites with more advanced detectors finally allowed Copernicus to be retired in 1981.

4.5 TD-1A

TD-1A was the first major project of the European Space Research Organisation and was a three axis stabilised satellite carrying a number of astronomical experiments. For

The TD-1A satellite, Europe's first satellite stabilised in 3 axes. (ESA).

its time it was a large and complex spacecraft and one which placed Europe on a par with much of the USA's space astronomy achievements. The name of the satellite reflects the planned Thor Delta launch vehicle and the fact that a second satellite (TD-1B), was originally proposed. Unfortunately, the programme ran into severe budgetary problems and was cancelled about one year into the C/D phase. The project was eventually rescued after a reorganisation of the ESRO financial structure, but there were insufficient funds for the second mission, and this never took place. The satellite was box-shaped, 1 m square by 2.16 m tall, with a pair of solar panels projecting on either side. The lower portion of the box was a service module containing attitude control, communications and power systems; the upper region contained seven scientific experiments all viewing out of the top. Attitude control was provided by momentum wheels and argon gas jets.

The TD-1A gamma ray and X-ray experiments have already been described (see sections 2.4.3 and 3.3), and only the two stellar ultraviolet experiments will be described here. Experiment S-59 was a spectrometer provided by the Space Research Laboratory of Utrecht University. This used a small telescope feeding a three slit scanner covering three wavelength bands between 210 and 280 nm. About 200 bright stars were observed by the instrument, an important result for the period, and these data were published by ESA in catalogue form.

Experiment S2/68, was an ultraviolet spectrophotometer, developed jointly by the UK and Belgium, which used an off-axis reflecting telescope to focus radiation onto a set of entrance slits. These in turn fed a photometer and a three channel spectrophotometer. The light falling on the spectophotometer entrance slit was reflected onto a diffraction grating and the dispersed light was then passed through one of three appropriately positioned slits and then onto individual photomultipliers. The orbital motion of the satellite caused the images of stars to move across the entrance slit which in turn caused the dispersed beam to scan across the exit slits without the need for any moving parts. The spectrometer operated in the range 133–263 nm, and, by reading out the counts in each photomultiplier every 148 ms, a resolution of about 3.5 nm could be obtained. The field of view of the S2/68 experiment was 17 arc minutes, so that the combination of spacecraft scan motion and the satellite's orbital precession of 4 arc minutes per orbit allowed the entire sky to be surveyed during the planned 6 month mission.

TD-1A was launched from Vandenberg Air Force Base on 11 March 1972 (12 March in Europe) and placed into near polar, Sun synchronous orbit at an altitude of about 540 km. The satellite pointed its solar panels at the Sun and then began to rotate slowly around its Sun pointing axis, sweeping out a great circle across the sky each orbit. This enabled TD-1A to carry out an ultraviolet survey, and, in the case of stars near the ecliptic pole which were observed on every orbit, to search for stellar variability. Data acquired by the experiments were transmitted to the ground when possible, the remainder being stored onboard and transmitted during ground station passes.

All went well at first, but by May both of the satellite's tape recorders had failed, and the only data that were being received were those taken when TD-1A was in direct contact with the ground. This amounted to only about 15% of the data taken by the satellite, so ESRO mounted a campaign to bring extra stations into use. Various CNES (French national space agency), NASA, and other ground stations volunteered their services, which improved the situation slightly (about 20–25% data recovery), but the scientists pressed for still greater efforts. Accordingly ESRO set up temporary ground stations in places as far afield as Singapore, Fiji, Tahiti, Easter Island, Marambio (in the Antarctic), and on the MS *Candide*, a 500 tonne banana boat stationed in the roaring forties, and managed to increase the data recovery rate to about 60%.

After six months in orbit, TD-1A entered a period of regular eclipses as the satellite passed behind the Earth and sunlight to the solar panels was cut off. This marked the end of the nominal mission, and controllers placed the satellite in a spin stabilised hibernation mode for four months. After the eclipse period ended the satellite, and its temporary ground station network, was reactivated, and a further six months of observations carried out. Despite the problems encountered in recovering data, the TD-1A produced a number of important ultraviolet catalogues.

4.6 OTHER ULTRAVIOLET EXPERIMENTS

About the time that Copernicus and TD-1A were operating, a number of other ultraviolet instruments were launched. Some were of a survey nature, others were targeted at more specific objectives, and each will be briefly mentioned here.

4.6.1 Apollo 16

During the Apollo 16 mission in 1972 astronauts John Young and Charles Duke deployed an ultraviolet camera/spectrograph on the lunar surface and used it to record images of the Sun, the Earth's geocorona, and various astronomical targets. The instrument used a 10 cm Schmidt camera, with a circular field of view 20° in diameter, to bring light onto a potassium bromide photocathode which was sensitive to photons with wavelengths shorter than 160 nm. When struck by a suitable photon the photocathode emitted electrons which were then accelerated by electric fields onto an electron sensitive film. Although apparently rather complicated, this 'electronographic' camera was about 20 times more sensitive than a similarly sized instrument which simply focused ultraviolet radiation directly onto a piece of film. The electronographic camera had the additional advantage that it was almost completely insensitive to normal light, which was not energetic enough to activate the photocathode, and so was able to view faint ultraviolet emission without needing filters to block out longer wavelengths.

The camera could be operated with two different corrector plates (thin lenses at the front of a Schmidt camera which reduce distortion), each of which had different characteristics. One plate, made of lithium flouride, transmitted wavelengths as short as 105 nm and thus included the important spectral line of hydrogen at 121.6 nm (Lyman alpha); the other had a short wavelength limit of 125 nm and transmitted a number of lines of oxygen and nitrogen. Comparison of images taken with the different corrector plates allowed the relative contributions of Lyman alpha and other emissions to be assessed.

In its spectrographic mode the camera was rotated so that instead of viewing the sky directly it pointed downwards. In this position the light reaching the camera had first to pass through a slit and then be reflected back upwards from a diffraction grating (1200 grooves/nm) mounted below the camera, producing a spectrum of the objects in the field of view. The wavelength range of the instrument in this mode was about

The Apollo 16 ultraviolet camera deployed on the Moon. (NASA).

50–100 nm, limits set by the reflectivity of the coatings on the diffraction grating.

To operate the instrument the astronauts removed it from the lunar module and set up the camera in the shadow of the lunar module, using the shadow direction and the position of the Earth in the sky as reference points. An astronaut then pointed the experiment in the required direction and triggered an automatic sequencer which made a series of direct and spectroscopic images. There was no drive mechanism to follow the stars as the Moon turned on its axis, but the effect of trailing is much smaller on the Moon because of its slower rotation.

The instrument operated successfully, and images of the Earth and its surroundings, as well as ten different regions of the sky, were obtained. At the end of the observations the film was removed from the camera and returned to Earth for analysis. As well as obtaining scientific data, the experiment also demonstrated the value of electronographic cameras for ultraviolet astronomy, and established the practicality of using the lunar surface as a site for astronomical observations.

4.6.2 Skylab

The American Skylab space station was launched in 1973 and occupied on three occasions during 1973–4. Neither the mission nor the extensive set of solar instruments carried on Skylab will be described here, but three ultraviolet experiments carried inside the space station will be mentioned. For further details of the Skylab mission see the *Bibliography*.

Astronaut Alan Bean operates the S019 ultraviolet experiment at the scientific airlock of the Skylab space station. (NASA).

Experiment S019 used a 15 cm telescope converted into a spectrograph by placing a calcium fluoride prism in front of it. The experiment was operated by the astronauts, who were required to place the spectrograph in a small scientific airlock in the side of the space station, extend and adjust an articulated mirror to view the required region of sky, and then advance the film and operate the camera shutter manually. The instrument had a field of view of 4° by 5° and was operated on all three Skylab missions. A total of 188 star fields, covering approximately 3660 square degrees, were photographed. Over 1600 stars were detected at wavelengths shorter than 200 nm. Some images were smeared out by motions of Skylab during the exposure, and these were subsequently computer processed to recover as much useful data as possible.

Experiment S201 was an electronographic ultraviolet camera similiar to the one used on Apollo 16. It was carried on the last manned visit to Skylab with the intention of observing Comet Kohoutek. Like S019, the instrument was deployed in the scientific airlock and viewed the sky via the mirror system. As well as observations of the comet, other objects including the Pleiades star cluster, the Gum Nebula, and the Small Magellanic Cloud were photographed.

Experiment S183 was developed by French astronomers from experiments carried on sounding rockets. The objective was to obtain ultraviolet intensities of hot stars and galaxies in three wavelength bands. To do this the instrument used two separate systems, a two channel photometer and a Schmidt–Cassegrain telescope. Both systems were mounted in the same housing and viewed through the scientific airlock via the articulated mirror system. Light from an area about 7° by 9° was directed into the photometer via the mirror and focused onto a diffraction grating. Two spectral bands, centred on 188 nm and 297 nm, were isolated from the dispersed beam by a mask and each then passed through an array of lenses and onto a photographic plate. The lenses

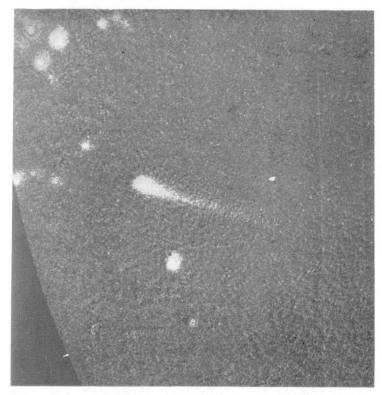

Comet Kohoutek recorded by the S201 Ultraviolet experiment on the Skylab space station (NASA).

spread out the light from each star into a rectangular image which could be compared with an image produced by a calibration source. The use of these smeared images avoided difficulties in interpretation caused by overexposure of the central regions of a well focused bright star. The Schmidt–Cassegrain telescope recorded the sources in its 5° by 7° field of view by using a 16 mm camera. The desired wavelength region, which lay between the two bands of the photometer, was selected by using coatings of suitable reflectivities on the telescope mirrors. Thirty-six star fields were observed by the instrument.

4.6.3 Orion 1 and 2

The Orion 2 instrument was an ultraviolet spectrograph carried onboard the Soviet Soyuz 13 mission in December 1973. The experiment was mounted on the outside of the Soyuz orbital module and operated by cosmonauts Pyotr Klimuk and Valentin Lebedev from a console inside the spacecraft. The spectrograph had a field of view of 20° square and was pointed by first aligning the entire Soyuz with the target and then adjusting the pointing with the instrument's own motors and star trackers, providing an accuracy of a few arc seconds. During the eight day mission spectrograms of over 3000 stars, some as faint as 11th magnitude, were obtained. The film was removed from the instrument and returned to Earth at the end of the mission. The instrument itself was destroyed when the discarded orbital module of the Soyuz spacecraft burned up on re-entry.

A similar instrument, Orion 1, had been carried inside the pressurised compartment of the Salyut-1 space station. Unfortunately, cosmonauts Georgi Dobrovolski, Victor Patseyev, and Vladislav Volkov died during the return to Earth on 29 June 1971 when their Soyuz capsule was accidentally depressurised after undocking from Salyut-1. Professor Grigor Gurazadian, designer of the Orion experiments, stated that the Salyut-1 crew had proved the concept of the Orion 1 instrument, and that useful data were obtained from film recovered from the Soyuz 11 capsule.

4.6.4 Astronomische Nederlandse Satelliet (ANS)

The ANS, launched on 30 August 1974, was the first Dutch national satellite. The spacecraft, its X-ray instruments and the launch and mission operations are described in section 2.4.3. The ANS carried a photometer to measure the brightness of stars in five wavelength bands with a view to extending the photometric classification scheme used for hot stars into the ultraviolet. The instrument used a 22 cm Ritchy–Chretien telescope to bring light onto a curved diffraction grating (2400 lines/mm) located below the primary mirror. The dispersed beam was chopped into five bands by a set of masks in the focal plane of the grating, and the five separate beams were then re-imaged via a set of lenses onto individual photomultipliers. The photometric bands covered the range from 156–329 nm and were each about 15 nm wide.

The satellite was pointed at selected stars, using information from the attitude control system together with data from star sensors in the telescope itself. The brightness of the star in the various wavelength bands was then recorded before moving onto the next target. ANS operated for 12 hours under onboard control, and during

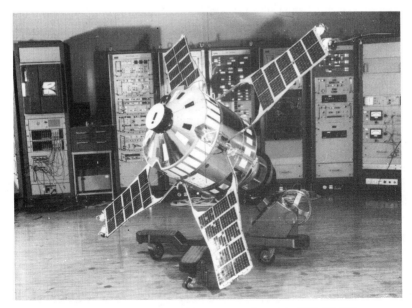

The French D2B-AURA satellite before its launch in 1975. (CNES).

this time would locate, lock onto, and observe up to 40 stars before transmitting the recorded data to the ground and recieving new instructions.

4.6.5 D2B-AURA

The D2B satellite was the third in a series of French satellites designed for launch by the Diamant rocket. The spin stabilised satellite was a cylinder 0.7 m in diameter and 0.8 m high to which were attached four short solar panels. Attitude control was provided by cold gas jets. The satellite carried four ultraviolet experiments for both solar and stellar observations, hence the designation of the satellite as AURA—Astronomical Ultraviolet Radiation Analysis. D2B-AURA was placed into an approximately circular, 400 km orbit inclined at 37° on 27 September 1975. The instruments studied both point sources and diffuse ultraviolet emission such as the zodiacal light, the Gegenschien or counterglow, and the ultraviolet emission from the Earth's atmosphere.

A similar satellite, but carrying gamma ray detectors in place of the ultraviolet instruments, was launched as part of the Franco–Soviet SIGNE programme. This was known as D2B-Gamma.

4.6.6 The Galactika experiment on Prognoz-6

The Soviet Prognoz-6 satellite, launched into a highly elliptical orbit in September 1977, carried a Franco–Soviet ultraviolet spectrophotometer called 'Galactika'. The instrument, which covered the wavelength range 110–190 nm in a number of 20 nm bands, looked in the direction away from the Sun to record the ultraviolet background in regions 6° square.

Ultraviolet radiation from the Earth's geocorona was found to be considerable at low altitudes and to extend to about 100 000 km. Fortunately the high apogee of the satellite (200 000 km) allowed observations to be made from above this background, and some 4000 scans covering regions such as bright areas of the Milky Way, the Pleiades star cluster, and the bright B star zeta Taurus. It was found that once away from the Milky Way the regions where the diffuse ultraviolet emission was highest were those known to be regions of enhanced soft X-ray brightness. This correlation suggests the existence of a hot component of the interstellar medium, possibly connected with supernova remnants, in these regions.

4.7 THE INTERNATIONAL UTRAVIOLET EXPLORER (IUE)

The IUE is one of the most successful space astronomy missions ever launched. It exceeded its design life by many years, proved the concept of a space observatory operated by visiting astronomers, and produced data for hundreds of scientific papers, yet very nearly didn't get off the ground at all.

IUE began life as the Large Astronomical Satellite, a 1964 UK proposal for an ESRO mission which was axed on financial grounds in 1967. It re-emerged as the Ultraviolet Astronomical Satellite, but was axed again. Eventually Professor R Wilson, who had been a prime mover behind the proposal, turned to NASA. Despite the problem of convincing the Americans to develop a proposal twice rejected by Europe, the NASA Goddard Spaceflight Center were persuaded to take up the idea, and, for a time at least, the mission was considered for inclusion in the Small Astronomy Satellite series as SAS-D. Europe eventually rediscovered its interest in the proposal, and by the early 1970s the mission had evolved into a tripartite project between NASA, ESRO, and the UK.

Under this agreement NASA provided the satellite, most of its scientific payload, and one ground station; ESRO (later ESA) provided solar panels and a European ground station; and the UK provided the sensitive ultraviolet cameras required for the scientific instruments. It was also agreed that the satellite would be operated like a ground observatory, with astronomers participating in their observations from the satellite control centres, and that observing time was to be divided between the three agencies so that the USA recieved two thirds of the time, with the UK and ESRO sharing the remaining one third. It was also agreed that observing time should be open to astronomers from around the world, hence the final designation of the satellite as the 'International Ultraviolet Explorer'.

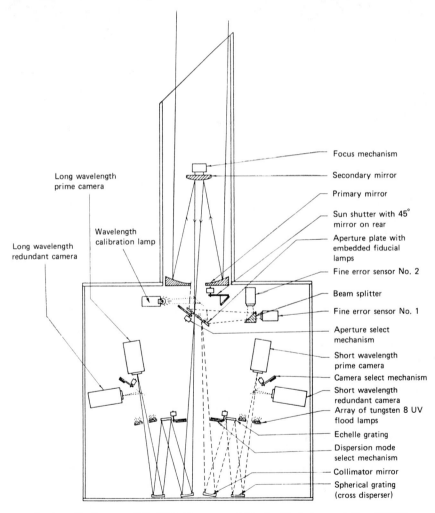

A schematic diagram of the IUE telescope and its scientific instrumentation. (SERC).

From its geosynchronous orbit IUE can remain in contact with the NASA Goddard Spaceflight Centre around the clock. Long periods of communications with the ESA station at Villafranca (Spain) are also possible. (ESA).

The IUE spacecraft consists of an octagonal body with various electronic systems arranged around the inside walls. A central, cylindrical, core down the centre of the body contains the telescope. The telescope projects from one end of the body, and a boost motor, required to transfer the satellite into its final orbit, protrudes from the other. The IUE is 4.2 m long and weighs about 700 kg. Angled solar panels extend from each side of the body near the bottom. Attitude information is provided by a combination of gyroscopes, star sensors in the scientific instrument, and a Sun sensor. The satellite is pointed by means of momentum wheels which can be unloaded by using hydrazine thrusters. The thrusters are also used to maintain the satellite in the correct orbit, a procedure known as station keeping.

To operate as an observatory the IUE must be in continuous contact with a ground station, and to achieve this the orbit chosen was geosynchronous[†], a long ellipse stretching between 26 000 and 45 000 km above the Earth and inclined at 28.4° to the Equator. This orbit takes 24 hours to complete, and during this time the Earth rotates once below the satellite. The combination of these two motions means that from the ground IUE appears to wander slowly around an ellipse on the sky centred roughly over South America. The satellite is always visible from the USA, and is also visible from the European ground station near Madrid for about 12 hours per day. This means that American controllers can direct the satellite for 16 hours a day, handing over to

[†]The orbit is not geostationary since the satellite does not appear to remain at the same point in the sky.

The IUE telescope before installation in the satellite. (SERC).

European controllers for the remaining 8 hours. This achieves the planned division of observing time without the need for astronomers to fly back and forth between Europe and the USA.

The IUE was conceived as a mission devoted to spectroscopy, so the scientific payload consists of the telescope, a pair of spectrographs, and a set of ultraviolet sensitive TV cameras. Fine error sensors—star trackers which help keep the telescope pointing in the required direction—are included as part of the scientific instrument. The telescope is a 45 cm $f/15$ Cassegrain configuration with a Ritchey–Chretien figure which provides a uniform image shape across the 16 arc minute field of view. The primary mirror is made from beryllium, the secondary mirror from fused silica, and the telescope tube from aluminium wrapped in Mylar thermal insulation. The telescope is fitted with a sunshade angled at 43° to the optical axis, which prevents sunlight entering the instrument provided that the IUE always points in a direction greater than 43° from the Sun. The 43° angle was chosen to allow observations of Venus when the planet was at its greatest angular distance from the Sun.

In the focal plane is a mirror, angled at 45° to the telescope axis, in which are drilled small holes. These holes act as entrance apertures for the spectrographs. Most of the light entering the telescope is reflected off this mirror onto the fine error sensors which relay an image of the field of view to the ground. Astronomers use this image to confirm that the telescope is pointing in the correct direction. The sensors then measure the position of the target star relative to the spectrograph apertures, and the IUE is manoeuvred so that light from the target is sent down the appropriate aperture. The fine error sensors then use a star near to the target to guide the telescope and ensure that it remains accurately pointed during the observation.

The two spectrographs operate over different wavelength ranges, 115–195 nm and 190–320 nm respectively. Each has two entrance apertures, a hole which appears 3 arc seconds in diameter and an oval slot 10 × 20 arc seconds in size. The slots can be closed by mechanical shutters if required. Light entering the spectrograph falls on an Echelle diffraction grating and is dispersed and reflected onto a second grating which disperses the beam further and reflects the spectrum onto one of the ultraviolet cameras. Movable mirrors placed in the beam divert the light to one of two cameras in each spectrograph. A second moving mirror can be inserted between the two diffraction gratings to reflect the beam onto a camera after only one dispersion. This produces a spectrum with a resolution of 0.6 nm (compared with 0.02 nm in high resolution mode), but allows fainter objects to be observed. A lamp situated close to the focal plane is used both for wavelength calibration and to flood the cameras with light to remove all traces of the previous image.

The cameras operate by first converting ultraviolet photons to electrons with a caesium—tellurium photocathode. The electrons are then accelerated away from the photocathode by a powerful electric field and strike a phosphor screen where each electron stimulates the production of about 60 optical photons. The image produced on the phosphor screen is then transferred, via a fibre optic link, to a television camera which builds up an image electronically as it stares at the phosphor screen. The camera can build up an image over many hours, making long observations of very faint sources possible.

Before each observation, controllers send commands to erase all trace of the previous observation and to prepare uniform, low noise background level as a baseline for the next exposure. The object under study is then observed and a spectrum is built up in the ultraviolet camera. At the end of the exposure the spectrum recorded on the camera tube is read, a destructive process by which the charges stored on the camera are read off, digitised, and transmitted to Earth. If for any reason the data are lost in transmission, the exposure must be started again since the information stored on the camera tube is wiped off during the readout process.

The data received at the ground station are transmitted to the control room and displayed to allow the astronomers to check their results and modify the observing programme if required. All the data are stored, and after the observing session is finished are computer processed; calibrations and other corrections are applied to produce a final, calibrated, dataset. The visiting astronomer can then take away the data, usually stored on magnetic tape, for analysis. All the data are archived, and after an appropriate time, at present 6 months after the final data tape is delivered to the observer, the data are made available to any interested astronomers who apply to the archive.

IUE was launched on 26 January 1978, and, after separating from its Delta launcher, it used its own motor to reach its planned elliptical orbit. Slight inaccuracies in the orbit cause the satellite to drift slowly in longitude, but this can be corrected by firing thrusters a few times each year. Since launch the satellite has operated almost perfectly, and long ago exceeded its three year design life.

The IUE's long life is due to the provision of backup systems and the ingenuity of its control team. For example, each spectrograph contains two cameras, which is fortunate because one camera in each spectrograph has developed serious faults since launch. The availability of spare cameras has meant that the scientific capability of the satellite has not been compromised by these failures. Another example, which illustrates the skills of the ground team, concerns the gyroscopes in the attitude control system. In theory three gyroscopes are required to provide the information to point the satellite, but as a precaution IUE was fitted with six, arranged in such a way that any three could be used. One gyroscope ceased operating in early in 1979, and over the next few years two others failed, leaving just the minimum of three working. Then, in August 1985, one of these also failed. Undaunted, ground controllers developed methods of operating with the two remaining units plus information from the Sun sensor, and plans to control the satellite with just one gyroscope plus the Sun sensor have been prepared against the day that another gyroscope fails.

The longevity of IUE has been important because, as well as allowing many more observations than expected, it has been possible to detect long term changes in variable sources and to make observations of transient phenomena like comets and novae. So great have been the rewards from the mission that it is difficult to find an area of astronomy that has not benefitted from results from IUE, and only brief highlights can be mentioned here. For more details see the Bibliography.

In nine years IUE observed 26 comets, including Enke's comet which was studied during two different trips around the Sun. An important result of this has been measurements of the emission from hydroxyl (OH) groups in a number of comets. Since the hydroxyl groups come from the breakdown of water molecules, this information, combined with ground based studies at other wavelengths (for example in the infrared) has allowed the relative proportions of gas to dust to be determined in many different comets. IUE also detected sulphur in the bright comet IRAS–Araki–Alcock and monitored Halley's comet during its 1985/6 appearance. The satellite has also been used for observations of asteroids, to monitor Jupiter's atmosphere, and to study the cloud of ions associated with Jupiter's moon, Io.

Beyond the solar system IUE has revolutionised our understanding of stars. It has confirmed observations from Copernicus that massive, and hence hot, stars have powerful stellar winds—streams of ions much denser and faster than the gentle outflow from the Sun. Although the amount of material in the winds is small, over hundreds of thousands of years this steady loss of mass may have profound effects on the evolution of these stars. Cool stars have also proved interesting when observed in the ultraviolet. Although a cool star does not emit much ultraviolet from its visible surface (the photosphere), there is considerable emission, mostly in the form of spectral lines, from its upper atmosphere. This arises in the transition region where the hot chromosphere merges into the very hot, and X-ray emitting, corona. Like X-ray observations, IUE data are helping to unravel the complex processes in which energy is transported from the cool photosphere, through the chromosphere and into the million degree corona.

Another example of the IUE's flexibility has been the study of novae. Since novae cannot be predicted, astronomers must wait until one occurs and then reschedule the IUE observing programme to make the required observations. In this way more than ten novae have been observed, and the data obtained have improved our knowledge of these stellar outbursts. It now seems clear that novae occur in binary star systems containing a giant star with a white dwarf companion. Material from the giant star spills over onto the white dwarf until a critical concentration of material is reached on the white dwarf's surface. When this happens the built up layer explodes, producing the temporary, large increase in brightness which characterises a nova.

An even more remarkable tribute to the flexibility of the IUE observatory came with the discovery of Supernova 1987A in the Large Magellanic Cloud during the early hours of 24 February. Within hours of the discovery IUE broke into its normal programme to begin a series of regular observations to record the ultraviolet spectrum of the supernova as it faded. The IUE observations are important for two reasons; firstly, the supernova explosion itself is of great interest, and secondly, by providing a bright source of ultraviolet light the supernova enables astronomers to search for absorption lines caused by gas in the long line of sight to the supernova.

Although of tremendous interest, especially since observations began so soon after the initial discovery, the IUE observations did cause some confusion during the first hectic attempts to identify which star had exploded. The position of the supernova suggested that the progenitor was a blue star called Sanduleak $-69°202$, but less than a week after the explosion the short wavelength ultraviolet flux faded dramatically and the light that remained closely resembled the ultraviolet spectrum of a star very like Sanduleak $-69°202$, the star which supposedly no longer existed. Ground based astronomers remeasured the positions of the supernova and rejected the idea that the supernova might have been a star close to Sanduleak $-69°202$, and so teams of IUE researchers returned to examine their data more closely. After some study the IUE team confirmed that the original interpretation of the data had been incorrect; the faint ultraviolet light detected after the supernova faded had come from two nearby blue stars, and Sanduleak $-69°202$ was indeed the progenitor of the supernova.

IUE has detected ultraviolet emission from normal galaxies and from exotic objects such as Seyfert galaxies and quasars. One discovery is that the emission from one Seyfert galaxy (NGC 4151) consists of continuum radiation from an energetic central object and emission lines from hot gas clouds in orbit around the central powerhouse. Detailed monitoring by IUE has shown that both continuum and line emission vary, but not simultaneously. The time lag is due to the distances between the central region and the orbiting clouds; it takes time for the changes at the centre to affect the clouds where the line emission is produced. Since the energy travels outwards at the speed of light, the time lag allows the distance of the clouds from the centre to be calculated. Furthermore, since the emission lines are broadened by the Doppler effect as the clouds orbit the centre of NGC 4151, it is possible to calculate their orbital velocities. Once both the orbital velocity of the clouds and their distances from the centre of the galaxy are known, it is possible to use Newton's laws to work out the mass of the central powerhouse. This turns out to be so large that it indicates the presence of a supermassive black hole at the centre of NGC 4151, and by implication in many other active galaxies as well.

Some research programmes carried out by IUE.

Studies of the chemistry of over two dozen different comets.
Attempts to determine mineralogical compositions of asteroids by reflectance spectroscopy.
Investigations of the chemical composition of the atmospheres of the giant planets Jupiter, Saturn, Uranus, and Neptune.
Observations of the moons of Jupiter and Saturn and of the Io torus.
Observations of Pre-Main-Sequence stars such as T Tauri and Herbig Ae/Be variables.
Studies of the chromospheres of cool stars.
Observations of the powerful stellar winds from hot massive stars such as those of spectral types O, B, and Wolf–Rayet stars.
Investigations of cataclysmic variables (novae) including the mechanisms of accretion in these unusual binary systems.
Studies of white dwarf stars and planetary nebulae.
The development of an ultraviolet spectral classification system for hot stars.
Determination of the chemical composition of the interstellar medium, both locally (that is, within a few hundred parsecs) and throughout the Galaxy.
Studies of supernovae and supernova remnants.
Observations of globular clusters.
Determinations of elemental abundances in the Magellanic Clouds.
Studies of active galaxies and quasars, including estimating the mass of the black hole believed to lie at the centre of the Seyfert galaxy NGC4151.

4.8 ASTRON 1

The Astron 1 mission, launched 23 March 1983, is a Franco–Soviet ultraviolet space observatory. The 4 tonne spacecraft is based on the Venera interplanetary craft, but is fitted with astronomical experiments in place of the Venus entry capsule. The main instrument, which projects several metres from the spacecraft body, is an $f/10$, 80 cm telescope, developed under the direction of the Crimean Astrophysical Observatory. The telescope has a field of view of about 30 arc minutes and is equipped with a three channel spectrometer built by the Space Astronomy Laboratory at Marseilles. It has an operating range of 150–350 nm with a spectral resolution of a few nm. Light entering the spectrometer is dispersed by a diffraction grating, and the resulting spectrum is scanned by a pair of photomultipliers mounted on a precision carriage able to move in very small steps.

A typical observation begins with the Astron spacecraft being pointed in the required direction and then tilting the telescope's secondary mirror to steer the light from the object of interest into the spectrometer via one of three diaphragms in the telescope's focal plane. The diaphragms are of different sizes so that objects with a range of brightness and angular extent can be studied. Like IUE, the satellite then uses a nearby star as a reference point to stabilise itself during the observation.

A set of small X-ray instruments known as the SKR-O2M experiment are also carried on the spacecraft. This experiment, which has a field of view of about 3°, covers the energy range 2–20 keV.

The Astron-1 satellite was placed into a highly eccentric orbit (1950 × 201 000 km) with a period of 98 hours. In principle this should allow long observations, but in practice communication sessions are limited to 5-6 hours of which about half is actually spent observing the objects of interest. The orbit also keeps the satellite well clear of the Earth's radiation belts and allows observations to be made without interference from ultraviolet emission produced in the Earth's tenuous outer atmosphere. The satellite was still operating at the end of 1987, by which time it had observed numerous stars and galaxies in detail, as well as making special observations of Supernova 1987A in the Large Magellanic Cloud.

4.9 SPACELAB 1

Two ultraviolet experiments were carried on the first Spacelab mission, launched on 28 November 1983, but both were less successful than hoped. An ultraviolet camera, known as FAUST, covered the range 170–220 nm and was planned to record objects as faint as about magnitude 17.5. Unfortunately, when the film was developed it was found to have been badly fogged. This fogging was originally ascribed to ultraviolet emission from the Shuttle itself, but later analysis suggested that it may have been due to intense emission of spectral lines at 130 and 135 nm produced by ionised oxygen in the upper atmosphere.

The second instrument, also an ultraviolet camera, was deployed through a scientific airlock in the Spacelab module. The instrument had a 66° field of view and was sensitive to the 130–300 nm range. Unfortunately, this experiment also suffered from fogged film, but in this case it is believed that sunlight reached the film and fogged it directly. More than half of the 48 pictures taken were spoiled. A further disappointment was a mechanical problem which prevented any spectrograms being made; only direct ultraviolet images were possible.

4.10 THE GLAZAR TELESCOPE ON MIR

The Glazar, a contraction of Galaxy and Quasar, telescope was installed in the Kvant astrophysics module which was docked to the Soviet Mir space station in 1987 (see section 2.5.5). The instrument is an ultraviolet telescope with an aperture of about 40 cm. It can be pointed with an accuracy of about 2 arc seconds, using its own fine pointing system. The telescope, which has a field of view of about 1.3°, records an electronically intensified image onto film which is replaced at intervals by the cosmonauts on the Mir station. The film is then returned to Earth for analysis. The object of the experiment is to conduct a survey of Markarian galaxies, which are known to be unusually bright in the ultraviolet, and to observe clusters of stars within our Galaxy. A second Glazar telescope may be fitted to Mir at a later date. If the Kvant module is still attached when the second telescope arrives, the new instrument will be mounted so that it is pointed at right angles to the unit on Kvant.

4.11 FUTURE MISSIONS

4.11.1 The Hubble Space Telescope

As yet there are no firm plans to launch a successor to IUE, although ultraviolet observations will be possible with the Hubble Space Telescope (see section 5.1).

Although primarily regarded as an optical instrument, the Hubble Telescope optics will have a short wavelength limit of 115 nm, about the same as that of IUE. The Hubble Telescope mirror is much larger than that of IUE and this, combined with improvements in the technology of ultraviolet detectors over the last decade, should make it more sensitive in the ultraviolet than its predecessors. However, the low orbit of the Hubble Telescope will mean that most of the observations which it makes will be much shorter than those possible with IUE, and this may make the two satellites roughly equivalent in the ultraviolet.

4.11.2 ASTRO-1

The ASTRO package is a set of three ultraviolet telescopes planned for flight in the Space Shuttle payload bay. All three instruments will be mounted on the ESA instrument pointing system and aligned so that simultaneous observations are possible. Astronomer-astronauts will operate the package by acquiring targets and guiding the telescopes via TV monitor displays in the Shuttle cabin. A typical seven day mission would provide about 60 hours of good quality observing time on the night side of the Earth with uninterrupted pointings of up to about 30 minutes being possible. Additional observations could be made during the sunlit parts of the orbit, but such observations would be subject to various pointing constraints.

The Wisconsin Ultraviolet Photo-Polarimetry Experiment (known by the almost unbelievable acronym of WUPPE) is an 0.5 m $f/10$ Cassegrain telescope which can measure the intensity and polarisation of objects as faint as 16th magnitude. These

If launched on time, ASTRO-1 would have observed Halley's comet (NASA).

measurements can be made simultaneously in a number of 4 nm bands over the range 140–330 nm. A high (0.5 nm) resolution mode is also available. The Ultraviolet Imaging Telescope (UIT) is a 38 cm, $f/9$ telescope with two, selectable, electronographic cameras. One camera operates from 120–170 nm, the other from 125–300 nm, and each is equipped with a six position filter wheel. The UIT is expected to detect 25th magnitude stars in a single, 30 minute, exposure. The largest instrument is the 90 cm, $f/2$ Hopkins Ultraviolet Telescope (HUT) which is designed for spectroscopy in the 45–200 nm range. The HUT uses a curved diffraction grating (600 lines/mm) to disperse the incoming beam and a linear array of 1024 photodiodes as a detector. The photodiode array is read out electronically and can provide a spectral resolution of 1.5–3 nm depending on the wavelength range being studied.

ASTRO-1 was scheduled to fly in early 1986 so that observations of Halley's comet could be made at about the same time that various spaceprobes intercepted the comet. It was to have been followed by at least two other missions in 1986/7, but all these missions have been delayed after the loss of the Space Shuttle Challenger. One mission is presently scheduled; the other two have been postponed indefinitly. One consolation for the long launch delay is that a broadband X-ray telescope will be added to the ASTRO-1 payload to allow studies of supernova 1987A in the Large Magellanic Cloud.

4.11.3 LYMAN

The success of IUE and Copernicus showed the desirability of extending ultraviolet observations to wavelengths shorter than those accessible from the Hubble Telescope. Although no such missions have yet been approved, a number of studies have been carried out. These include an ESA proposal for a satellite known as Magellan and the NASA Far Ultraviolet Spectroscopic Explorer (FUSE) proposal. Since both these missions had similar scientific objectives, it was agreed to study the possibility of a single cooperative venture. This potential mission was tentatively known as Columbus, but is now referred to as LYMAN.

As originally conceived, LYMAN would have explored wavelengths between 90 and 120 nm, a region which contains many important spectral lines and which cannot be studied with IUE. Secondary objectives were to observe at shorter (EUV) wavelengths and in the normal ultraviolet. LYMAN would have used grazing incidence optics, thus avoiding the difficult task of producing normal mirrors with adequate reflectivities at short wavelengths and at the same time providing the possibility of EUV observations. The main instruments would have been high resolution spectrographs. LYMAN would probably have been placed in low Earth orbit by the Space Shuttle. Unfortunately, the disruption to NASA's programme after the Challenger accident has reduced the chances of a joint mission taking place, and ESA has been forced to consider developing a smaller satellite capable of being launched by the Ariane rocket. This has thrown everything back into the melting pot, and, while LYMAN will doubtless eventually re-emerge, it is not yet clear what form the new proposal will take.

4.12 FAR ULTRAVIOLET ASTRONOMY

The region from about 50–90 nm is often called the far ultraviolet. Observations at these wavelengths are hampered by both the absorption of far-ultraviolet photons in the interstellar medium and by the difficulty of manufacturing suitable instruments. In particular the designer of far-ultraviolet telescopes faces two severe problems; the sensitivity of normal mirrors falls very quickly at wavelengths shorter than 120 nm, so telescopes relying on reflecting optics are inefficient, but grazing incidence telescopes working at these wavelengths are heavy and expensive to manufacture. For both of these reasons there has been little support for a satellite optimised for far-UV astronomy, although studies such as FUSE and LYMAN have often included detectors able to operate in this wavelength range as secondary objectives.

The only sustained programme of space based far-ultraviolet astronomy has been conducted with instruments on the American Voyager spacecraft. The Voyagers, whose main mission is to explore the giant planets and their satellites, each carry an ultraviolet spectrometer operating between 50 and 170 nm which is designed to probe the chemical composition of planetary atmospheres by recording ultraviolet spectra during planetary encounters. The spectrometers are relatively simple devices which use a reflective diffraction grating to disperse incoming light and image it onto a detector. The spectral resolution is about 2.5 nm.

During the long cruises between planets the Voyager spectrometers have been used to observe a number of hot stars, both nearby white dwarfs and more distant O and B stars, with a view to estimating the amount of hydrogen gas between these stars and the solar system. Most of the measurements have been made with the Voyager 1 instrument, since this spacecraft has completed its main mission. The observations have been successful, detecting several far-UV sources discovered from rockets and during the Apollo–Soyuz mission (see section 4.13.1), and discovering far ultraviolet emission from a small number of white dwarfs.

4.13 EXTREME ULTRAVIOLET (EUV/XUV) ASTRONOMY

Wavelengths between 6 nm and 50 nm cover the region known as the extreme ultraviolet. It was believed for many years that photoelectric absorption by interstellar gas (notably neutral hydrogen and neutral and singly ionised helium) would restrict the range of EUV observations to objects within a few parsecs of the Sun, but results from TD-1A and Copernicus revealed that the interstellar medium in the solar neighbourhood was much less dense than originally thought. This discovery implied that in certain directions the EUV horizon could be up to several hundred parsecs away and that EUV emission from a few stars might be detectable. Sounding rockets carrying EUV experiments were launched in the early 1970s, but the most important breakthrough came with the flight of an EUV telescope on the American half of the Apollo–Soyuz Test Project (ASTP) mission in 1975.

4.13.1 The ASTP EUV telescope

The ASTP instrument used a nest of four grazing incidence mirrors and was fitted with two selectable channel electron multiplier detectors and a six position filter wheel. The mirrors, the largest of which was 37 cm in diameter, were fabricated from nickel coated aluminium overcoated with a thin layer of gold. The telescope field of view was either 2.5° or 4.3°, depending on which detector was selected, and was sensitive to wavelengths between about 5 and 150 nm. The filter wheel had five filters with

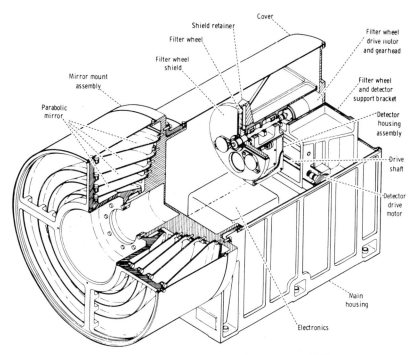

A schematic diagram of the ASTP EUV telescope (NASA).

passbands between 5 and 45 nm wide; the sixth position contained an opaque screen and was used to enable the background reading in the detector to be recorded. The instrument was mounted in the service module of the Apollo spacecraft and was operated by the astronauts in the command module. The Apollo attitude control system was used to orient the spacecraft so that the telescope pointed in the desired direction, and observations lasting between 1 and 20 minutes were then made. A typical sequence consisted of a period observing a possible EUV source and an equal time observing areas a few degrees away to measure the EUV background in the same general area. A significantly different count rate between the regions implied EUV emission from the chosen source. The instrument operated normally throughout the mission, and a total of 30 objects, including the planet Jupiter, were observed.

The main result was the discovery of several point sources of EUV radiation. Brightest of these was the white dwarf HZ43 (No. 43 in the catalogue of Humason & Zwicky), a very hot, and hence bluish, white dwarf with a faint red companion. HZ43 is about 60 parsecs from the Sun and is of particular interest since it appears that the emission from the white dwarf actually peaks in the EUV. HZ43 could thus be regarded as an EUV star which also happens to be visible at other wavelengths. A second hot white dwarf, known as Fiege 24, was also detected. The value of these observations is that comparison of the measured EUV emission with theoretical models enables our understanding of the final phases of stellar evolution to be tested.

Two other discrete sources of EUV radiation were detected, the stars Proxima Centauri and SS Cygni. Proxima Centauri, the nearest star to the Sun, is a red dwarf with a photospheric temperature of about 3000 K, much too cool to produce EUV radiation. However, it is a flare star (see section 2.4.3), and since giant stellar flares would produce considerable EUV radiation, it is likely that the ASTP telescope detected just such a flare. SS Cygni is a dwarf nova system, a binary containing a white dwarf accreting material from a companion star, which brightens by 3–4 magnitudes roughly every 50 days. SS Cygni was detected on three occasions during the Apollo–Soyuz mission at a time when ground based observers reported that the star was undergoing one of its semi-regular outbursts. The remaining 26 objects searched for were not detected, and in some cases this allowed upper limits to be placed on their EUV fluxes. An interesting consequence of this was that the white dwarf companion to Sirius, which was searched for but not detected, seems to emit much less EUV radiation than predicted.

The ASTP telescope also made a study of the EUV background, a diffuse flux at energies just below that of the soft X-ray background described in section 2.4.5. The instrument recorded data in its 5.5–15 nm band both near to preselected targets and also as it swept across the sky between pointings. This showed that the irregular structure of the soft X-ray background may not be mirrored by the diffuse EUV emission, but since the ASTP data covers only limited regions of the sky, confirmation of this conclusion must await the first true EUV sky surveys.

4.13.2 Future missions

The success of the ASTP instrument showed the potential value of a survey at EUV wavelengths, and two such missions are planned: the NASA Extreme Ultraviolet Explorer (EUVE) and the UK Wide Field Camera, an EUV telescope carried on the German ROSAT satellite (see section 2.6). Both missions were planned for launch by the Space Shuttle in about 1988, but have now been transferred to expandable launch vehicles and delayed until about 1991.

The basic concept of the EUVE is of a slowly spinning satellite equipped with four telescopes covering the EUV band. Three of the telescopes will point at right angles to the spin axis; the other will look along the axis. Since the spin axis will lie along the Earth–Sun line throughout the mission, the three outward looking telescopes will scan great circles across the sky and build up a complete survey in six months. The fourth telescope, looking down the spin axis but away from the Sun, will repeatedly observe a strip of sky along the ecliptic plane and build up a more sensitive survey of this limited region. At the end of the six month survey, a period of pointed observations will be carried out, using a spectrometer fitted to the telescope looking along the axis.

The EUV telescopes will be of the grazing incidence type, about 40 cm in diameter, manufactured from aluminium and then gold coated for maximum reflectivity. The detectors will use photocathodes that will produce electrons when struck by a photon and microchannel plates, similar to those used in X-ray telescopes, to amplify the original electron into a signal large enough to be detected. This two stage system was chosen so that information about where on the detector the original photon arrived will be preserved, making possible the production of images of the EUV sky. Each telescope will be fitted with a single filter with a bandpass of 20–40 nm. Between them, the four telescopes will cover the region from about 8–75 nm.

The spectrometer will perform moderate resolution spectroscopy in the 8–80 nm band during the pointed phase of the mission. It uses a set of three diffraction gratings arranged so that each intercepts some of the beam from the on axis telescope (the remainder of the beam falls onto the detector used during the survey phase). Each grating disperses the radiation falling on it and reflects the beam onto its own microchannel plate detector. The three gratings operate over different wavelength bands and will produce spectra with a resolution of 0.1–0.2 nm. Each detector is fitted with a broadband filter to stop unwanted ultraviolet photons being detected and contaminating the spectrum.

The configuration of the EUV Explorer has changed several times since the mission was conceived. The spacecraft was originally envisaged as a slowly spinning drum with four short solar panels projecting from its base. In this version the nozzle of a propulsion system, used to raise the orbital altitude to 550 km after a Space Shuttle launch, projected from the centre of the drum on the Sun facing side, and the deep survey/spectrometer telescope looked out of the opposite face. The three survey telescopes observed through holes in the walls of the cylindrical body at different points around the circumference.

This design was revised when it was decided to use the NASA Multimission Modular Spacecraft (MMS) to carry the EUVE payload. The MMS, first used on the Solar Maximum Mission, is a standardised satellite platform which can be launched and retrieved by the Space Shuttle. An MMS can be fitted with a variety of different payload modules depending on the objectives of a particular mission, and before the loss of the Space Shuttle Challenger there was a possibility that the EUVE might have been fitted to the already orbiting Solar Maximum Mission (SMM). In this case the SMM payload module would have been removed, and replaced with a platform carrying the EUVE experiments, by astronauts working in the Shuttle payload bay. After a careful checkout, the EUVE would then have been released to commence its mission. At the end of the EUV survey the spacecraft would have been retrieved by the Space Shuttle, its experiments replaced by another payload module, and the SMM redeployed.

In 1987 it was decided to transfer the EUVE to an unmanned launcher, probably a Delta rocket. The SMM is compatible with the Delta, so it will be possible to use a new SMM to carry the EUVE payload, and the process of in-orbit recovery and instrument changeover will probably be exercised to replace the EUVE payload with the next Explorer mission, the X-ray Timing Explorer (XTE).

The British-built WFC is an EUV telescope which forms part of the ROSAT mission. It is a cylindrical instrument approximately 1 m long and 60 cm in diameter attached to the ROSAT spacecraft so that its optical axis is co-aligned with the main X-ray telescope. The WFC uses a set of three, nested, Wolther–Schwartzchild type 1 grazing incidence telescopes to image a region of the sky 5° in diameter onto a caesium iodide photocathode. The electrons emitted when the photocathode is struck by EUV photons are amplified by a microchannel plate and read out electronically when they strike a position sensitive resistive sheet at the bottom of the microchannel plate. The WFC carries two identical detectors, one of which is held in reserve in case the other fails, and is expected to locate sources with an accuracy of one arc second or better. Various filters, mounted in a wheel, can be placed above the detector to define several broad wavelength bands. Data recorded by the WFC are passed to ROSAT for storage and subsequent transmission to the ground.

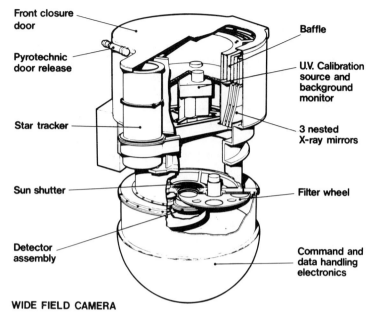

Diagram of the ROSAT Wide Field Camera. (SERC).

During the ROSAT survey the WFC will map the entire sky in two broad and overlapping wavelength bands between 6 and 20 nm to produce a comprehensive survey. During the second part of the mission, in which the satellite carries out detailed observations of preselected targets, the WFC will use additional broadband filters to extend its wavelength coverage to 70 nm. Combining pointed observation data from the main telescope and the WFC will provide considerable wavelength coverage of soft X-ray sources which are not so distant that their EUV radiation is absorbed by the interstellar medium.

Some highlights of space ultraviolet astronomy

1920s Attempts to observe the Sun from above the ozone layer, using balloons.
1946   Ultraviolet spectrum of the Sun obtained, using V2 rocket.
1957   Rocket experiments make first ultraviolet measurements of stars.
1961   First ultraviolet observations of stars in southern hemisphere by rockets launched from Woomera.
1964   Ultraviolet photometer carried on US military satellite 1964–83C. UK proposes satellite for ultraviolet observations. After repeated rejections this mission emerges as the IUE observatory.
1965   First UV spectra of stars good enough to resolve individual spectral lines are obtained during a rocket flight.
1966   OAO-1 fails shortly after reaching orbit.
1968   OAO-2 launched, operates beyond one-year design life.

1970 OAO-B destroyed during launch failure.
1972 OAO-3 (Copernicus) begins long series of UV observations. European TD-1A satellite launched to make UV sky survey. NASA and UK agree to develop International Ultraviolet Explorer.
1975 White dwarf HZ43 and a few other objects detected in extreme ultraviolet by an EUV telescope on the Apollo–Soyuz mission.
1978 International Ultraviolet Explorer launched.
1987 IUE used to study supernova 1987A in Large Magellanic Cloud within hours of its discovery.
1988 IUE celebrates ten years of operations; it is the longest lived astronomical satellite ever launched.

5

Optical astronomy from space

5.1 THE HUBBLE SPACE TELESCOPE

5.1.1 Introduction

Although the optical was the first, and for two thousand years the only, wavelength region used for astronomy it will be almost the last to be studied from space. The reason for this is that the atmosphere is transparent to normal light, so, unlike the unexplored spectral regions, there is little chance of any fundamentally new discoveries being made by a small optical telescope in space. The purpose of placing optical telescopes in orbit is not to open up new wavelength regions, but to escape the degradation of images caused by atmospheric turbulence. Freed from the effects of the atmosphere, an orbiting telescope can produce sharper images than its ground based counterparts, and can approach the ultimate goal of optical performance limited only by the effects of diffraction. Unfortunately, to make significant gains in angular resolution over ground based instruments, a space telescope must be large and must have its optical components made and aligned with very great precision, requirements which imply a heavy, complex, and thus expensive satellite. To make such a satellite cost effective requires that it must remain in operation for many years, and this has delayed the development of such an observatory until the era of the Space Shuttle.

The Space Telescope is a massive undertaking with a long and complex history, and because the programme is described in detail elsewhere only a brief summary of the satellite and its development will be given here. Interested readers are referred to the *Bibliography* for more details.

The first NASA sponsored studies of a large orbiting optical telescope were in 1962 and 1965, and these were followed by a series of seminars in 1967 and 1968 in which the idea was promoted throughout the astronomical community. In 1971 and 1972 NASA carried out studies of an instrument with a primary mirror 3 m in diameter which it referred to as the Large Space Telescope. Between 1973 and 1976 a scientific definition of the telescope, its instruments, and possible modes of operation was made by a group led by Dr C R O'Dell, and during this study the proposed size of the main mirror was cut to 2.4 m to reduce costs. About the same time the word 'Large' was dropped from the name of the proposed instrument.

European astronomers were anxious to join the project, and it was agreed in 1977 that ESA would develop the Space Telescope's solar panels and provide one of the scientific instruments, in exchange for which Europe would receive 15% of the observing time available. The Space Telescope project was formally announced in December 1976 and was approved by the US Congress in 1977. Several major industrial corporations expressed interest in the development work, with Kodak and Perkin–Elmer competing to manufacture the telescope, while Lockheed, Martin-Marietta, and Boeing sought the contract for the spacecraft systems. The final awards were to Perkin–Elmer and Lockheed.

Work on the optical systems began in 1977 with the casting of the blank for the primary mirror and its delivery to Perkin–Elmer in 1978. Grinding and polishing of the blank took three and a half years, ending in 1981. The mirror's reflective coating was applied in just four minutes on 5 December 1981. Despite being stored in ultraclean conditions, a thin layer of very fine dust built up on the mirror over the next three years, but this was eventually removed without affecting the mirror's optical performance. In 1983 NASA decided that the satellite would be named the Edwin P. Hubble Space Telescope, in honour of the astronomer who first showed that the Universe was expanding. This is generally contracted to Hubble Space Telescope, or HST.

The primary mirror of the Hubble Space Telescope before it received its reflective coating. Note the cellular structure of the lightweight core fused between the faceplates of Corning ultra-low-expansion glass. (Perkin–Elmer).

The optical systems, together with the scientific experiments and elements of the satellite structure, were delivered to the Lockheed plant at Sunnyvale in California in early 1985, and over a six week period that summer the various elements were joined together for the first time. Tests of the entire satellite then commenced, culminating in early 1987 with a 57 day thermal vacuum test during which the conditions which the instrument will face in orbit were simulated, and tests of the entire payload were carried out. These tests revealed some problems, including a shortage of electrical power, and efforts to correct these shortcomings began at the conclusion of the test programme.

The planned launch of the HST in late 1986 was disrupted by the loss of the Space Shuttle Challenger, so instead of being transferred to Cape Canaveral in August 1986 for loading aboard the Shuttle Atlantis, the satellite remained at Sunnyvale in storage. Four of the scientific instruments were removed during this period for minor modifications. The HST remained in storage until the resumption of Space Shuttle flights when it will received high priority, for a launch on one of the first few flights of the Shuttle.

Assembly of the Hubble Space Telescope at Lockheed's Plant at Sunnyvale, California. (Lockheed).

5.1.2 The HST spacecraft

The HST consists of three main elements: the telescope itself, the Support Systems Module with its solar panels, and the Scientific Instruments Module.

The $f/24$ telescope, known as the Optical Telescope Assembly or OTA, lies at the heart of the satellite. It consists of a 2.4 m primary mirror and a 0.3 m secondary mirror arranged in a Ritchey–Chretien optical system. The primary mirror, made of a fused glass honeycomb structure sandwiched between two sheets of ultra-low expansion titanium silicate glass, has a surface accurate to 0.000 025 mm and is coated with a thin layer of highly reflective aluminium and a protective layer of magnesium fluoride. This combination of coatings is expected to make the HST sensitive to wavelengths between 115 nm and 1 μm. The main structure is fabricated from a graphite epoxy composite which combines high rigidity with low thermal expansion and lightness. Special optical control sensors, located in the focal plane, allow ground controllers to monitor the condition of the telescope, and other sensors allow the curvature of the primary mirror to be checked. Any distortion of the mirror which occurs after the HST reaches orbit can be corrected by 24 actuators attached to the back of the mirror. Additional motors on the support structure can be used to move the secondary mirror to eliminate any misalignments that might develop after launch.

The support systems module (SSM) supplies the services required to operate the spacecraft. It is a collar which fits around the telescope structure about level with the

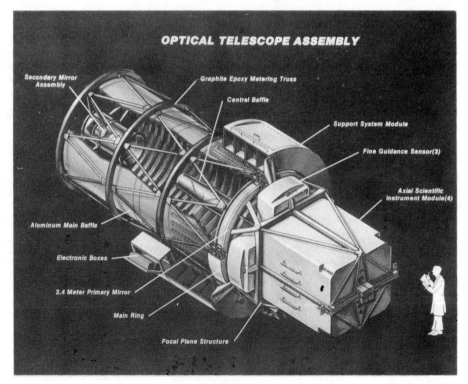

Simplified diagram of the Hubble Space Telescope. Note that mechanical systems such as the deployable solar panels and various parts of the outer structure are not shown. (Perkin–Elmer).

The Hubble Space Telescope

An impression of the Hubble Space Telescope in orbit. (Lockheed).

primary mirror and contains systems for electrical power regulation, communications, etc. The attitude control system, which must stabilise the telescope with an accuracy of 0.007 arc seconds, uses a combination of gyroscopes, star trackers, and fine error sensors to provide pointing information. The star trackers can determine the telescope pointing direction within a few arc minutes, which is sufficient to place preselected guide stars in the field of view of the telescope. The fine error sensors, grouped around the main mirror, use interferometry to provide error signals for the final pointing. A set of momentum wheels, which at maximum torque are able to slew the satellite through 90° in less than 20 minutes, is used to carry out all the manoeuvres.

The solar panel array, manufactured by British Aerospace, consists of two identical deployable and retractable solar arrays with a total area of about 50 sq metres. The arrays, mounted on a flexible substrate, will deliver about 4 kW of power after two years in orbit. Special emphasis has been placed on the dynamic stability of the panels in view of the very demanding pointing requirements of the HST.

5.1.3 The scientific instruments

The HST has five scientific instruments, each located behind the primary mirror in an instrument module. Four modules are mounted axially; that is, with their long axis parallel to the telescope, with a fifth mounted radially. The four axial modules are all the same size, and all five modules are designed to be independent of each other and capable of operation without mutual interference. This allows instruments to be changed in orbit during routine servicing, keeping the HST at the forefront of astronomical technology.

The Wide Field/Planetary Camera (WF/PC) is the instrument in the radial module. It can be operated in two ways loosely characterised by its dual name. The Wide Field mode will be used to survey selected areas for very faint objects, and the Planetary Camera mode will be used for imaging of smaller areas at increased angular resolution. Light entering the instrument is diverted by a movable mirror to either the Wide Field or Planetary Camera detectors, which each consist of four sets of 800×800 element electronic detectors known as charge coupled devices (CCD). The detectors, which are cooled to 177 K for increased efficiency, are coated with the organic phosphor coronene which increases their wavelength coverage by converting ultraviolet photons into visible light capable of being registered by the CCDs. The instrument is sensitive to wavelengths from about 115 nm to 1 μm and will detect objects over a range of magnitudes from 9–28. The minimum exposure time will be 0.1 seconds, with typical long exposures being about 3000 seconds, that is, half an orbital period. The WF/PC will be equipped with a variety of filters, transmission gratings, and polarisers, making it suitable for a wide range of observations.

The Faint Object Spectrograph (FOS) is designed for moderate and low resolution spectroscopy, spectropolarimetry, and time resolved spectroscopy. The incoming beam is dispersed by a diffraction grating and sent to one of two digicon sensors, one sensitive to ultraviolet and blue light, the other sensitive to visible and near infrared radiation. The digicons operate by the photoelectric effect, photons falling on the photocathode of the digicon produce a shower of electrons which are then focused by magnetic fields onto individual elements of a linear 512 photodiode. Since the electrons fall on the photodiode at a position which depends on where they emerged from the cathode, and the number of electrons emerging at each point is proportional to the number of photons reaching a specific area of the photocathode, the readout of the photodiode array can be electronically reconstructed to produce a spectrum. The limiting magnitude of the FOS will vary with wavelength and spectral resolution, but is expected to be about 21st magnitude at moderate resolution and 25th magnitude at low resolution, assuming a 3000 second exposure. For time resolved spectroscopy the instrument takes a series of spectra, each with durations of between 50 microseconds and 10 milliseconds, at rates of up to 100 spectra per second. The duration of each exposure will be set by the brightness of the source.

The High Resolution Spectrograph (HRS) can provide spectra with a resolution ($\lambda/\text{delta}\lambda$) of 100 000 at wavelengths between 110 and 320 nm. A moderate resolution mode is available for target acquisition, estimation of exposure times, and for observations at short wavelengths where the efficiency of the telescope is comparatively poor and a low resolution mode ($\lambda/\text{delta}\lambda = 2000$) will also be available. Like the FOS, the HRS uses two digicon detectors each with an array of 512 photodiodes. The limiting magnitude of the system will vary with wavelength, resolution mode, and exposure time, but for a 2000 second integration is expected to be about 11 at high resolution, 14 at medium resolution, and 17 at low resolution.

The High Speed Photometer (HSP) will be used to study brightness fluctuations over a wide spectral range and is capable of resolving events a few microseconds apart, a feat impossible from the ground because of atmospheric effects. The detectors are four image dissectors, essentially photomultipliers in which spatial resolution is provided by magnetic focusing, and a red sensitive photomultiplier. Two of the image dissectors operate at ultraviolet wavelengths and two in the visible and near ultraviolet. In combination with the red sensitive photomultiplier, this allows the spectral range from about 120 nm to 800 nm to be covered. Unlike a conventional photometer, the HSP contains no moving parts. Instead of rotating wheels used to position various filters and aperture stops in the incoming beam, the filter/aperture combination is set by positioning the image under study in a selected portion of the field of view. This positioning is done with small movements of the main telescope. The photometer will operate in one of three modes. The first is star/sky photometry with a single filter, a mode which allows the brightness of any local starfield or nebulosity to be subtracted to improve accuracy. The second mode is multi-band photometry in which filters are selected sequentially by small movements of the telescope. The third is wide field, filterless photometry over an area about 10 arc seconds in diameter.

The fifth instrument is the European supplied Faint Object Camera (FOC) which uses the full optical performance of the telescope to study very faint objects at high angular resolution. The FOC consists of two separate camera systems, operating at $f/96$ and $f/48$, which produce angular resolutions of 0.022 and 0.044 arc seconds respectively. At short wavelengths, even higher angular resolution can be achieved by inserting a compact Cassegrain assembly into the optical path to provide an $f/288$ mode. The detectors are based on the image photon counting system, essentially a very high performance image intensifier able to count individual photons, which is well suited to the study of very faint objects. With very long integration times (up to 10 hours) the FOC is expected to achieve reasonable signal to noise ratios on objects as faint as magnitude 28. The FOC is equipped with a variety of filters, polarisers, etc., and in the $f/96$ mode an occulting coronograph can be used to suppress stray light from bright objects when observing faint sources in the same field. The $f/48$ mode provides a long slit spectrographic mode for observing extended objects such as comets, nebulae and galaxies.

The HST can also use its fine error sensors to make positional (astrometric) measurements. With the aid of neutral density filters, the sensors should be able to observe objects of magnitudes between about 4 and 20. Angular distances between a number of objects in the same field of view should be measurable, with high precision, in a few minutes. As well as stars, possible targets for astrometric measurements include distant comets and the satellites of the outer planets.

5.1.4 Operating the HST

The 11.6 tonne HST will be carried into a orbit by the Space Shuttle and lifted from the Shuttle by the Remote Manipulator System. After a preliminary checkout by ground controllers, the HST will be released and the Shuttle will retire to a safe distance while the HST deploys its solar arrays and undergoes further checks. If all is well the Shuttle will return to Earth; if not, the telescope may be retrieved for repairs.

Immediately after deployment the HST will begin a one month orbital verification phase during which the telescope will be activated and focused, and engineering details such as the performance of the solar panels will be checked. This will be followed by a five month scientific verification phase during which the correct operation of the various instruments is confirmed and high priority astronomical observations are

Deployment of the Hubble Space Telescope from the Space Shuttle. (Lockheed).

carried out. At the conclusion of the scientific verification, the HST will be declared operational, and the first round of routine scientific observations will commence.

At this point scientific management of the HST will be passed to the Space Telescope Science Institute (STScI), a special facility set up by the Association of Universities for Research in Astronomy (AURA). AURA is a consortium of American universities which oversees the United States' national optical observatories, and has set up the STScI to operate the HST as another national facility. The STScI is located on the campus of the Johns Hopkins University in Baltimore (conveniently close to the HST operations centre at the Goddard Spaceflight Center), and is directed by Riccardo Giacconi, whose ability to manage complex astronomical space missions was proved during his work in X-ray astronomy. The STScI has a staff of over 250, including about 60 PhD astronomers, divided up into groups with diverse functions such as instrument calibration, guide star selection, operations and data management. The main function of the STScI is to provide facilities and advice to users of the telescope, but the STScI staff will also be expected to carry out their own research programmes.

After allowances have been made for engineering operations, and time has been allocated to the groups responsible for building the scientific instruments, the remainder of the observing time will be allocated to other astronomers. A small amount of observing time will be held in reserve for special projects and to allow flexibility to react to targets of opportunity such as novae or unusual and unexpected comets. In 1987 it was announced that time will also be allocated to suitably qualified amateur astronomers.

Communications with the HST will be via the Tracking and Data Relay Satellite System (TDRSS), a pair of satellites in geostationary orbit which can communicate with satellites orbiting below them. Since they provide a link between low flying satellites and ground controllers, the TDRSS satellites avoid the communications problems usually encountered with satellites in low Earth orbit and allow the HST to be operated as an observatory. Unfortunately, although two way communication with the HST should in principle be possible for about 85% of the time, the demands of other users of the TDRSS system are expected to reduce the available communication time to about 20%, and this means that full real time control of the telescope will not be possible. Consequently, most HST observations will be preprogrammed, with a sequence of commands being stored onboard and then executed automatically. For observations of unusual or irregular sources a variety of preplanned observations will be stored, and the astronomers, who will visit the STScI during their observations, will be able to choose between these preplanned alternatives in real time. These types of observation will be scheduled in blocks when near continuous communications are possible.

At the conclusion of each observation the STScI will reduce the data into a form suitable for further analysis, which may be carried out at the Institute or at the observer's home institution, and then archive the observation. After one year this data will be released into the public domain and will be available to any astronomer who requests it. Exceptions to this rule will be made for various long term projects, for example measurements of stellar parallax which, by their nature, must last for much longer periods.

5.1.5 Servicing the HST

After a few years in orbit the HST will be visited by a Space Shuttle. After the satellite has been grappled by the Shuttle manipulator and mounted on a special platform in the payload bay, astronauts will carry out a range of servicing tasks and may replace one or more of the scientific instruments. At the conclusion of the servicing the HST will be released to continue its planned observing programme. It was originally planned that the HST would be returned to Earth for major refurbishment every five or six years before being relaunched for another extended period of observations, but this idea has now been abandoned. All servicing and repairs are now expected to be carried out in orbit.

The ability to replace scientific instruments is a key element of the HST programme, and, because of the time they take to develop, preparations for the next generation of instruments is already well in hand. Engineers at the NASA Jet Propulsion Laboratory are building a spare WF/PC, and ESA is expected to develop a reserve FOC in due course. Whether these instruments will be significantly improved versions, or just copies of the originals, will be determined on grounds of cost, potential scientific benefit, and technical risk. Totally new instruments, some of which will extend the operating range of the HST into the near infrared, are also being developed.

Two infrared instruments are being designed, although only one of them is expected to be installed during the first round of instrument changes. The Near Infrared Camera and Multi Object Spectrometer (NICMOS) will cover the range 0.8–2.5 μm and can be used for both imaging and spectroscopy. It will be equipped with infrared array detectors (see section 6.3.1) based on germanium (covering 0.8–1.6 μm) and mercury–cadmium–telluride (covering 1.6–2.5 μm), and will provide a spatial resolution of up to 0.05 arc seconds in its high resolution mode. When used as a spectrometer NICMOS will position fibre optic links at different points in its field of view to record spectra of up to four sources simultaneously at a resolution $\lambda/\text{delta}\lambda$ of up to 10 000. The NICMOS detectors will be cooled by a refrigerator containing solid methane and solid ammonia which is expected to operate for up to five years. The Hubble Imaging Michelson Spectrometer (HIMS) is the alternative infrared instrument and will operate over almost the same wavelength range as NICMOS. HIMS will use arrays of 128×128 mercury–cadmium–telluride detectors to simultaneously image and record spectra of all the objects in its 10 arc second square field of view. Like NICMOS, HIMS, will require a supply of coolants, and the current plan calls for sufficient for at least 1.5 years of operations.

A third possible instrument is the Space Telescope Imaging Spectrograph (STIS), designed to operate in the ultraviolet and visible range. The STIS is expected to exceed the capabilities of both of the HST's first generation spectrographs (FOS and LRS) and will have imaging capability, making it possible to examine many points in a single region simultaneously rather than observing one point at a time. The field of view of the STIS will be 50 arc seconds square, and in its highest resolution mode a resolution of

Astronauts examine the Hubble Space Telescope during servicing by the Space Shuttle. Note that in reality the telescope cover would be closed to prevent danger of contamination from thruster gases etc. (NASA).

100 000 should be possible. It is expected that the STIS will be capable of studying objects up to three magnitudes fainter than any other instrument on the telescope.

5.2 HIPPARCOS

HIPPARCOS is a European mission whose objective is summarised by the rather contrived acronym, HIgh Precision PARallax COllecting Satellite, a name which commemorates the Greek astronomer Hipparcos (190–120 BC), who is remembered both for his star catalogue and for his attempts to determine the distance from the Earth to the Moon. Hipparcos laid the foundations of astrometry, the accurate measurement of celestial positions and motions, and this branch of astronomy will be carried to a new level of precision by the satellite that bears his name.

Although astrometry is no longer regarded as one of the glamorous areas of astronomy, until the development of astrophysics in the 20th century, all astronomers made astrometric measurements. It was the precision of Tycho Brahe's observations of the supernova in 1572 which shattered the idea that the Universe beyond the Moon was unchanging, and his measurements of planetary positions which provided the data Kepler used to develop his theories of planetary motion. Edmund Halley worked with cometary positions to make his prediction that the comet of 1672 would return, and in the 19th century astrometry was used to determine the distances to the nearest stars. Despite the achievements of 20th century astrophysics, astrometry remains the only direct way of measuring the distance to individual stars, and hence is crucial to all astronomical distance measurements. The importance of these measurements is that an accurate knowledge of the distance scale enables the energy given off by astronomical

The principle of astronomical parallax measurements. (ESA).

objects to be calculated and thus sheds light on the underlying physical processes occurring in them.

The object of most astrometric measurements is the determination of parallax, the angular displacement in the apparent position of an object when observed from two separated points. In daily life the effect of parallax can be seen by holding a pencil at arm's length and then closing first one eye then the other; the pencil seems to move relative to objects in the background. The stars are so distant that to detect the apparent shift of even nearby examples it is necessary to use the effect of the Earth's annual journey around the Sun; observations taken six months apart are separated by the diameter of the Earth's orbit. Even with this tremendous baseline the apparent shift of the nearest stars is less than one second of arc. Although making sufficiently accurate measurements is difficult, once the parallax of an object is known, it is straightforward to use trigonometry to calculate its distance in terms of the diameter of the Earth's orbit. Trigonometric parallax is so important that the distance to most astronomical objects is often expressed in parsecs, a unit defined as the distance at which a object would show an annual parallax of one arc second when observed from the Earth. A parsec is equal to 3.26 light years or about 10^{13} kilometres.

The measurements of astronomers like Tycho Brahe were restricted in accuracy to the resolving power of the human eye, about 1 arc minute, but with the invention of the telescope the precision of astrometric measurements steadily increased. In 1838 F. W. Bessel succeeded in measuring the parallax of the star 61 Cygni, and thus made the first direct determination of the distance of a star from the Earth. Since Bessel's work, and that of several of his contemporaries, improvements in technology have allowed the parallax, and hence the distance, of a large number of stars to be determined. Astrometric observations have also allowed the proper motions (that is, movement across the sky due to the actual motion of a star through space) of nearby stars to be measured. These observations seem to indicate that some stars follow slightly sinuous paths across the sky, and this has been interpreted as being due to the gravitational effects of massive planets around these stars.

Unfortunately, the accuracy of ground based astrometry is limited by factors such as the flexure of telescopes due to gravity, limitations imposed by atmospheric seeing, and the fact that no one observatory can observe stars over the entire sky. Together, these effects mean that determinations of trigonometric parallax can be made only for stars quite close to the Sun and place a limit on the accuracy with which the proper motions of stars can be studied. A significant improvement in the accuracy of astrometric measurements could be used for a number of important projects, including extending the range of parallax measurements to more distant stars, thus improving our knowledge of the entire cosmic distance scale, and confirming the reported irregularities in the motions of nearby stars which may indicate the presence of planets.

It was to overcome the problems inherent in Earth based astrometric techniques that the idea of a space astrometry mission was proposed. In the mid 1960s Professor Pierre Lacroute, of the Strasbourg observatory, developed a method of making astrometric measurements from a satellite, and in 1974 the idea was put to ESA for evaluation. The proposal was selected for a phase A study, but the study revealed a number of technical uncertainties and the idea was rejected in favour of a less demanding mission. Work, however, continued throughout the late 1970s and eventually, in mid-1980, ESA began phase B studies of the concept. Contracts for construction and development of the satellite (phase C/D) were placed with the French company Matra in 1984.

The principle behind the Hipparcos satellite remains that suggested by Professor Lacroute: the use of a split mirror which simultaneously looks at two widely different regions of the sky and superimposes the two images on a single focal plane. If each of the two superimposed images includes a star, the angle between the two stars is equal to the angular distance of the star images in the Hipparcos telescope plus the angle between the two halves of the split mirror. Since the positions in the focal plane can be measured with great accuracy, and the angle of the split mirror is known, the angle between the two stars can be found. In isolation this single measurement is of limited value, but by using a satellite that sweeps across the sky, many millions of star pairs can be measured and the data numerically combined to produce a tight net of positions that covers the whole sky. In the case of a mission lasting several years, repeated scans of the sky can be used to determine both parallaxes and proper motions for thousands of stars.

The Hipparcos mission is designed to determine the positions of 100 000 preselected stars with a precision of 0.002 arc seconds, a staggeringly small angle roughly equal to that subtended by a man standing on the limb of the Moon when viewed from the Earth. To achieve this, Hipparcos uses a telescope with a mirror 290 mm in diameter, but which is split into two semicircular portions with an angle of 58° between them. The combined image falls onto a series of finely ruled grids in the focal plane and, as the satellite rotates, the image moves across the focal plane. As the image marches across the grids, detectors below the grids see the stars flash on and off many times a second. The detectors convert the flashing starlight into a varying electrical signal, and the data processing system uses the phases of the modulated

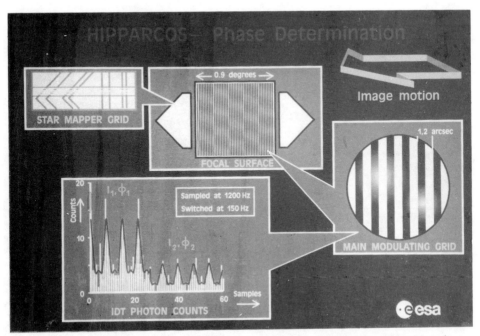

As stars seen by the Hipparcos telescope pass over these finely ruled grids detectors mounted below the grids see the stars flash on and off. Ground computers use this information to determine the relative positions of the stars which appear in the image. (ESA).

Optical astronomy from space

signals and the characteristics of the grid to determine the angular separation of the two stars. Since there is a strong chance that at any moment there will be more than two stars in the field of view, the detectors are Image Dissector Tubes (IDTs), basically photomultipliers which can restrict their attention to a single small region of the focal plane at any one time and thus avoid confusion.

To determine the pointing direction of the satellite, and thus enable the IDTs to look at the correct portion of the field of view, the focal plane contains a set of star trackers. As in the main instrument, these also consist of ruled grids, but in this case the layout is a combination of parallel lines plus a chevron pattern. As the image of a star moves across the star tracker grid the modulation of the signal is not regular, but is determined by where on the chevron pattern the image falls. From this, plus information from the spacecraft's own attitude control system, the pointing direction of the satellite can be found. The star tracker is so accurate that its data will be used to compile a second catalogue known as TYCHO. The TYCHO catalogue will contain the positions of about 400 000 stars, but at lower precision than the main experiment.

The optical system, which uses a flat mirror to fold the telescope beam and reduce the overall length of the instrument, is carried inside a hexagonal spacecraft 3 m tall and 2.5 m in diameter. Power is provided by three solar panels which are attached along alternate edges at the base of the hexagon and which swing outwards once the satellite reaches its operating orbit. Attitude control is provided by a hydrazine rocket system during the initial phases of the flight and by smaller cold gas jets once scientific observations have commenced. Hipparcos will be placed in a geostationary orbit by an Ariane rocket. As the satellite drifts towards its operating position at 12° West

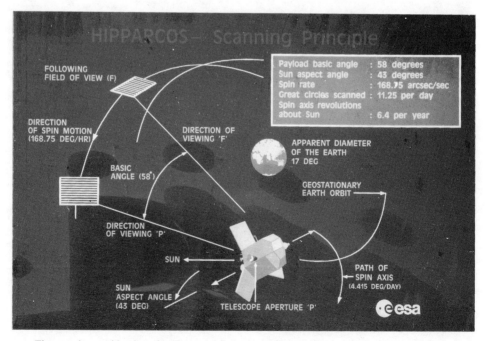

The complex combination of angles and rotations which allow Hipparcos to scan the entire sky repeatedly during its mission. (ESA).

longitude, a process which is expected to take three weeks, it will spin at 60 rpm but, since the spinning satellite will be slightly unstable, the hydrazine thrusters will be used to keep it within the required attitude limits. On reaching its operational location, the spin rate will be reduced to 10 rpm and any remaining hydrazine fuel will be dumped to prevent unwanted motions caused by excess fuel sloshing about in the tanks. The smaller cold gas system will then slow the satellite spin rate to 1 rpm, slew Hipparcos to the correct angle with respect to the Sun (spin axis inclined at 43° to the Sun vector), and reduce the spin rate to one rotation every two hours. Once these manoeuvres are complete, and the solar panels have been deployed, the scientific instrument will be activated and Hipparcos will begin its two and a half year mission.

Once Hipparcos is in its operational configuration the combination of the spacecraft's slow spin and the precession of the spin axis around the Sun vector means that the telescope will scan the entire sky repeatedly during the mission. As the satellite sweeps across the sky, its attitude will be measured by the star trackers, and the direction in which it is scanning will be compared to the planned one. If there is a difference of more than 10 arc minutes, the attitude control thrusters will be fired to make a correction. Since thruster firing disrupts the observations, the attitude control strategy requires that corrections are made in all three axes whenever any one axis moves out of alignment, even if the other axes are still within the allowed 10 arc minute error band. This procedure reduces the total number of periods during which thrusters are being fired, and minimises the loss of observing time.

Hipparcos will be operated from the ESA ground station at Darmstadt in W. Germany which will be in continuous contact with the satellite. Since the mission requires the measurement of preselected stars, and the rejection of all other objects in the field of view, the best available positions of the selected stars are stored in a computerised input catalogue at the ground station. Every few minutes ground controllers will use this catalogue, together with their knowledge of the pointing direction of the satellite, to prepare a list of the next stars to be measured. Details of when and where these stars will cross the focal plane of the telescope will be transmitted to the satellite so that that IDT detectors can be commanded to observe the correct portion of the focal plane. This procedure will ensure that the satellite measures the correct stars and ignores any others which happen to lie in the same field of view.

Data will be transmitted to the ground at 24 thousand bits per second throughout the mission. Once received the data will be checked and then sent to various experimenter groups for analysis. These groups will process the data to generate the two main data products, the Hipparcos catalogue and the larger, but lower accuracy, TYCHO catalogue. The immense amount of data processing required means that the catalogues will take some years to produce, and they will probably be released to the general astronomical community about 1995. After release the catalogues will be used for many kinds of individual astrometric studies and will form an astronomical reference system which will be used well into the 21st century.

As well as the specific scientific objectives of the Hipparcos mission, the fact that Hipparcos and the Space Telescope will be in operational simultaneously provides an opportunity for complementary astrometric measurements by the two satellites. Several of the HST instruments are capable of astrometry of very faint objects with an accuracy of a few milliarcseconds, but these measurements will be relative to other objects in the same field of view. In some cases these fields will include distant quasars,

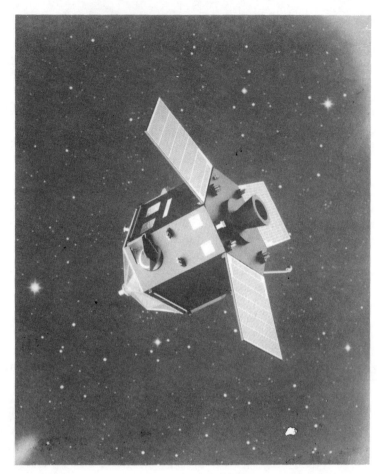

A model showing the appearance of Hipparcos in its operating orbit. (ESA).

which can be considered inertially fixed, and some fairly bright normal stars. Since Hipparcos will observe stars over the entire sky in the magnitude range from about 1 to 13, some of the stars which Hipparcos will measure will appear to lie close to quasars. Using the quasar positions as reference points, the Hipparcos and Space Telescope measurements can be linked together to provide a very rigid and consistent system of star coordinates without regional errors or any small residual rotation of the entire Hipparcos astrometric data set which might otherwise remain undetected. A further possible link between the two missions is that the HST requires a star catalogue of considerable accuracy to provide guide stars for use during observations. A catalogue based on ground based observations is being prepared by the STScI, but this will doubtless be updated by Hipparcos data during the operating life of the HST. Like its human predecessor, the latter day Hipparcos will leave a legacy of astrometric data which will be used by future generations of astronomers.

6

Infrared and millimetre astronomy from space

6.1 INTRODUCTION

The infrared is the extension of visible light to longer wavelengths, and can be regarded as covering wavelengths from 1 μm to 1 mm. The long wavelength end of the range is also known as the submillimetre region, which in turn overlaps with the radio wavelengths described in Chapter 7.

Infrared radiation was discovered by Sir William Herschel in about 1800, when he used a prism to split sunlight into a spectrum and placed a thermometer in the bands of different colours to investigate the relationship between the wavelength of light and its heating power. After trying each colour in turn, Herschel placed his thermometer beyond the red end of the spectrum and discovered that the Sun had some power outside the visible spectrum: Herschel thus made the first astronomical observation in the infrared.

Other infrared observations followed; Piazza Smyth detected infrared from the Moon in 1856, and a few years later the 4th Earl of Rosse made infrared measurements with his telescope at Birr Castle, Ireland. Rosse observed the Moon throughout its cycle of phases and attempted to derive the temperature of the lunar surface. In 1878, Thomas Edison recorded a solar eclipse and observed the star Arcturus, making what was probably the first infrared observation of an object outside the solar system. Unfortunately, the insensitivity of early detectors, and the absorption of infrared wavelengths by the atmosphere, meant that few astronomers showed an interest in infrared observations, and the new astronomy was virtually stillborn. Although infrared observations were made from a number of observatories between 1900 and 1910, the limitations of early instruments meant that infrared astronomy lay almost dormant until new electronic detectors were developed in the 1950s.

The first of these new detectors used lead sulphide (PbS) crystals. Radiation with wavelengths between about 1 and 4 μm, to which the atmosphere is at least partly transparent, can cause a physical change in a lead sulphide crystal, changing its

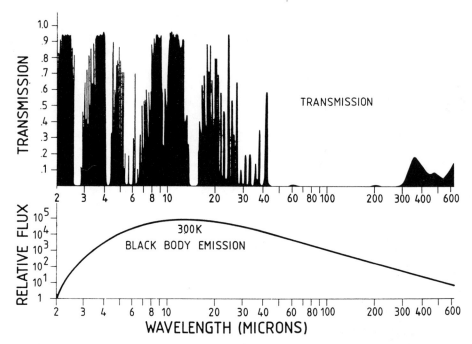

Diagram showing the transmission of infrared radiation through the Earth's atmosphere and the infrared emission from a black body at a temperature of 300 K (27 °C). Note the narrow atmospheric windows where transmission is large, and the lack of any transmission between about 40 and 350 μm. (ESA).

electrical resistance. Since resistance is easy to measure it is possible to use lead sulphide to make infrared detectors suitable for astronomy. With such instruments, cooled with solid carbon dioxide or liquid nitrogen to improve sensitivity, astronomers began to study the near infrared region for the first time. These observations gave a new perspective on the Universe because infrared radiation is emitted by material cooler than the surface of a star, and its longer wavelength enables it to penetrate clouds of dust that block the passage of visible light. Infrared detectors allowed astronomers to probe the dark interstellar clouds and reveal some of the secrets of starbirth.

To detect radiation beyond the 4 μm cutoff of the lead sulphide detector, and so study still cooler material, astronomers and physicists sought new infrared detectors. One idea was to use superconductors, materials which, if cooled below a critical 'transition temperature', have practically no electrical resistance. If a superconductor could be maintained just below its transition temperature and then exposed to infrared radiation, the energy falling on the material would cause its temperature to rise slightly, destroying the superconducting properties and producing a dramatic change in resistance. This method proved unworkable in practice, but it inspired physicists to try to use the strange half metals known as semiconductors to develop a detector working along similar lines.

In a normal metallic solid each atom loses control of some of its outer electrons, and these freed electrons wash around the material like a sea; electrical interactions between the electron sea and the atoms hold the metal together. Because the electron sea is free

The surveys

to move about within the material, the electrons respond easily to electric fields, and so metals are good conductors of electricity. Semiconductors are similar to metals, but their outer electrons remain weakly bound to the parent atoms and cannot escape into the material without some external stimulus. By introducing traces of elements which have either one more or one less outer electron than the semiconductor (a process known as doping) it is possible to adjust the properties of a semiconductor so that infrared photons provide the stimulus to start electrons moving about in the material. Since moving electrons constitute an electric current, which can be amplified and detected, a semiconductor can be used as the basis of a sensitive infrared detector. The first such detectors used germanium doped with copper (Ge:Cu) or with mercury (Ge:Hg) and were sensitive to photons from about 5 to 15 μm. Unfortunately this improved wavelength coverage came at a price; the new detectors would work only at very low temperatures, and astronomers found themselves cooling their instruments to 4.2 K by surrounding them with baths of liquid helium.

The Ge:Cu and Ge:Hg detectors were in turn superseded by a device developed by physicist Frank Low and based on germanium doped with gallium (Ga:Ga). Such a detector is equally sensitive to all infrared wavelengths and is called a bolometer. The Low bolometer made it possible for astronomers to observe at all the infrared wavelengths to which the atmosphere was transparent, but once again there was a price; bolometers must be cooled to less than 2 K. This is colder than the boiling point of liquid helium and can be reached only by continuously pumping gas out of the helium reservoir cooling the detector, thus lowering the pressure over the liquid and reducing its boiling point below the normal value of 4.2 K.

The difficulties of operating instruments at low temperatures, the narrow and often variable atmospheric windows available, and the powerful infrared emission from the atmosphere itself make infrared astronomy a very demanding science. These, and other, obstacles made the first two decades of ground based infrared observations a period of trial and error, but also a time of important new discoveries. The infancy of infrared astronomy need not concern us further except to note that the infrared was soon established as an important branch of astronomy.

6.2 THE SURVEYS

Satellites with infrared and submillimetre experiments are listed in Table 6.1.

6.2.1 The 2.2 μm and AFGL surveys

The first infrared sky survey was the 2.2 μm catalogue published by Professors Neugabauer and Leighton of the California Institute of Technology. Neugabauer and Leighton built their own telescope, an innovative instrument with a mirror produced by spinning a 1.5 m dish containing slow setting epoxy resin. As the dish span, the epoxy resin flowed into the shape of a parabola and then set, producing a crude but acceptable infrared telescope. Using this telescope and PbS detectors the astronomers set out to examine the 30 000 square degrees of sky visible from California. The project took six years and resulted in a catalogue, published in 1969, which contained 5612 sources.

About the time that the 2.2 μm survey was completed the US Air Force Cambridge Research Laboratories (also known as the Air Force Geophysics Laboratory or

Table 6.1.
Satellites with infrared and submillimetre experiments

Name	Launch date	Launch vehicle	Perigee (km)	Orbit[†] Apogee (km)	Inclination (°)	Notes
Salyut 4	26 Dec 1974	D-1	337	350	51.6	ITS-K IR telescope spectrometer (1–7 micron range) on manned Space Station.
Salyut-6 BST-1M	29 Sep 1977	D-1	380	390	51.6	Sub-mm telescope on manned space station
IRAS	26 Jan 1983	Delta 3910	896	913	99	US/Netherlands/UK. IR survey
Prognoz-9	1 Jul 1983	A-2e	380	720 000	65.5	RELIKT-1 Experiment
SL-2 IRT	29 Jul 1985	Space Shuttle (Challenger, STS-51F)	311	319	49.5	Spacelab-2 IR Telescope

[†] Since satellite orbits change because of atmospheric drag etc., orbital parameters quoted by different sources may vary

AFGL) decided to carry out an infrared survey using instruments carried above the atmosphere. The name of the project was HISTAR, and it used a 16.5 cm liquid helium cooled telescope carried on a sounding rocket. Seven launches were made from the White Sands missile range between April and December 1972, followed by two launches in 1974 from Australia to provide coverage of the southern hemisphere. In total about 30 minutes of observing time was accumulated from the nine flights. The AFGL survey covered about 90% of the sky in two broad wavelength bands centred on 11 and 20 μm. In addition, smaller regions of sky were covered at 4 μm (18%) and at 27 μm (34%). Full details of the survey were classified at the time, but a catalogue of over 2000 celestial sources, some but not all of which also featured in the 2.2 μm survey, was released in 1974. The AFGL continued with rocket experiments using larger telescopes and longer integration times (that is, observing a single region for a longer period) to detect fainter infrared sources throughout the 1970s, but the short observing periods available from rockets limited the amount of sky that could be observed. The time was ripe for the development of a satellite that could carry out a sensitive infrared survey of the entire sky.

6.2.2 The Infrared Astronomical Satellite (IRAS)

The original infrared sky survey satellite was based on the idea of a small spacecraft like the Uhuru X-ray mission, and was proposed to NASA in the mid-1970s. While the proposal was being considered, Dutch scientists revealed that, following the success of the ANS satellite (see section 2.4.3), the Netherlands were also studying a possible infrared satellite. With NASA still facing the budgetary crisis which had already affected the HEAO programme (see section 2.4.6) it was decided that the best chance for the mission was as an international project, and the scientists from the two countries combined to produce a joint proposal. Soon after the US–Dutch collaboration was agreed, the United Kingdom expressed an interest in the mission and joined the project, which became known as the Infrared Astronomical Satellite.

The main objective of the IRAS mission was to carry out a survey in broad wavelength bands centred on 12, 25, 60, and 100 μm. This survey was required to cover at least 95% of the sky with 99.8% reliability (that is, the number of spurious sources in the final catalogue would be less than 0.2% of the total number of entries) and to be reasonably complete over regions of the sky not confused by a very high density of sources (that is, apart from regions like the galactic plane where sources are very closely crowded together, only a few per cent of real sources would fail to appear in the catalogue). A secondary objective was to carry out detailed studies of selected objects with both the main survey detectors and a package of additional instruments.

Responsibility for various aspects of the mission were divided up between the three countries. The United States built a liquid helium cooled telescope and equipped it with an array of survey detectors. The telescope was fitted to a Dutch built spacecraft, and the satellite was controlled from a ground station in the United Kingdom. IRAS was launched by an American rocket, the Dutch provided the extra experiments, and the British and Dutch together developed the software needed to control the satellite. Scientific aspects of the mission were managed by a joint US/European Science team which together organised the production of the final infrared catalogue at the NASA Jet Propulsion Laboratory in California.

The development of IRAS was plagued by what Gerry Neugabauer, US chairman of the joint management committee, described as 'an infinite series of crises'. This period, and some of the many political and technical problems, is described in *The cosmic enquirers* (see *Bibliography*), and will not be recounted here. Inevitably, the problems meant that the launch was repeatedly delayed, and more than once the possibility of total cancellation hovered over IRAS as costs escalated. Fortunately, the international aspects of the mission made total cancellation politically impossible, and eventually, in January 1983, IRAS was ready for launch.

The key to the IRAS mission was that the sensitivity of an infrared telescope is determined not only by the detectors used, but also by the amount of unwanted infrared energy received from the sky and from the telescope itself. In the case of ground based instruments, both the atmosphere and the telescope structure are warm and so emit considerable infrared radiation of their own, swamping the weak infrared signals from the stars and limiting the achievable sensitivity. In space the effect of the atmosphere is removed, and it is possible to cool the entire telescope to very low temperatures, reducing the emission from the structure and greatly improving the overall sensitivity of the instrument. Accordingly, IRAS was built around a cryogenically cooled, 60 cm, $f/9.6$, reflecting telescope of Ritchey–Chretien configuration. The telescope had a field of view of about 30 arc minutes and used mirrors machined from beryllium in such a way that they assumed the correct shape when cooled. To maintain the required temperatures, the entire telescope was contained within a large, cylindrical tank which, at the time of launch, contained 72 kg of superfluid helium (that is liquid helium cooled to less than 2 K). The liquid helium, boiling slowly away as the mission progressed, cooled the focal plane to 2.5 K. Other parts of the telescope were slightly warmer, but all remained below 10 K throughout the mission.

To reduce the rate at which the helium coolant boiled away, IRAS was covered with multilayer thermal insulation, and the allowed pointing directions of the satellite were strictly controlled to prevent unnecessary heat reaching the telescope either from the Sun or from the Earth. A combination of a sunshade and a series of cooled baffles within the telescope prevented stray light reaching the focal plane. The telescope was topped by an ejectable aperture cover which protected it from contamination before launch and prevented material evolved from the satellite condensing on the cold optical surfaces.

The telescope focal plane was dominated by an array of 62 semiconductor detectors each sensitive to one of the four wavelength bands chosen for the IRAS survey. The detectors were laid out in an ingenious pattern such that every source crossing the field of view would be seen by at least two detectors in each wavelength band. This arrangement provided a means of eliminating spurious signals caused by charged particles hitting the detectors; any astronomical source seen by one detector should be seen by another similar detector a few seconds later. This automatic check was known as 'seconds confirmation'.

The focal plane also contained the apertures for the Dutch Additional experiment, or

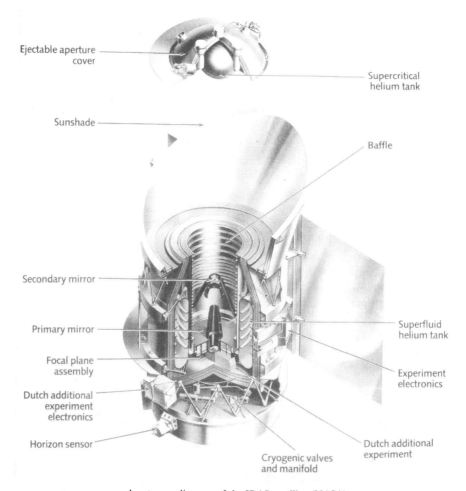

A cutaway diagram of the IRAS satellite. (NASA).

The surveys

The IRAS focal plane. The 62 infrared detectors appear as small rectanglar slots in the central portion and the slits for the star trackers are in the segments on either side. The scan direction was such that infrared sources crossed individual detectors across their short axis. Note how the detectors are staggered to assist in determining the positions of objects which pass across the focal plane. The largest detectors are 7.5 × 5 mm in size. (Ball Aerospace Corporation).

DAX. The DAX consisted of two separate elements; a Low Resolution Spectrometer (LRS) and a Chopped Photometric Channel (CPC). The LRS was a slitless spectrometer covering wavelengths from 8–13 and 11–23 μm with a resolution λ/deltaλ of about 30. It was triggered automatically whenever a bright source entered its field of view. The CPC operated in bands of 41–63 and 84–114 μm and was designed to map extended objects with a resolution of 1.2 arc minutes. The LRS was thus somewhat better for mapping than the main survey detectors which had an angular resolution of 4 arc minutes.

The remaining space in the focal plane was taken up by eight slit-like visible light detectors, four on either side of the main infrared array. Four of the sensors were arranged at right angles to the direction in which IRAS scanned; the others lay at angles of about 45° to the scan direction. A star crossing the field of view triggered the detectors on one side of the array, and, by timing the intervals between each detection,

The launch of IRAS from Vandenberg Air Force Base in California. (NASA).

the precise position of the star relative to the focal plane could be calculated. Combining this information with data from the satellite's attitude sensors, the pointing direction of the telescope could be determined with an accuracy of better than 20 arc seconds.

The orbit chosen for IRAS was circular, near polar and at an altitude of 900 km. The orbit was Sun synchronous with the satellite orbiting virtually above the boundary between day and night. Under normal circumstances IRAS was pointed so that it looked directly away from the Earth and surveyed a strip of sky 30 arc minutes wide every orbit. The gravitational effects of the Earth's equatorial bulge caused the orbit to precess by about 15 arc minutes (half the telescope's field of view) per orbit, so each region of sky was observed twice in quick succession. This repeat scan was used as a further check for the reliability of sources detected by the survey array; celestially fixed objects seen on one orbit were 'hours confirmed' on the next orbit 90 minutes later.

One additional check on the reliability of the sources detected was performed. Although IRAS always pointed more or less at right angles to the Sun, there were times when it was allowed to point slightly closer to the Sun than the nominal scan direction, and others when it lagged slightly behind. This provided an opportunity to rescan each region and to cross check detections with sources seen a week or so earlier. This process was known, logically enough, as 'weeks confirmation'. The cumulative effect of the 15 arc minutes per orbit precession, combined with the regular periods of looking back to carry out weeks confirming scans, amounted to a drift in scan direction of about 1° per day and allowed IRAS to observe the entire sky in about six months.

Although it might appear unnecessarily complicated, the purpose of the multiple confirmations was essential to ensure that the final IRAS catalogues did not contain spurious sources. Objects such as other satellites, asteroids, and comets are all warm enough to emit considerable infrared radiation and to produce false entries in the IRAS catalogues. Fortunately, almost all such objects move sufficiently to fail confirmation at either the seconds, hours, or weeks level and so could be rejected during final data processing. Asteroid and comet data were, however, extracted for publication in a specialist IRAS catalogue of solar system objects.

Although IRAS was primarily a survey mission, not all of the time could be spent in survey mode because of various operating constraints, so the remaining time, about 40%, was allocated to engineering checks, calibrations, and additional (pointed) observations. Each additional observation allowed about eight minutes of observing time on a single object before the rapid orbital motion of IRAS moved the target out of the region of sky over which observations were permitted. These observations were used to map extended sources or to make sensitive surveys of small regions of sky known to be of special interest. The additional observations were divided between astronomers from the three participating countries according to a formula agreed by an international team set up to manage the scientific aspects of the mission.

The IRAS spacecraft provided electrical power, attitude control and housed the telemetry systems and the onboard computer. Attitude control was provided by

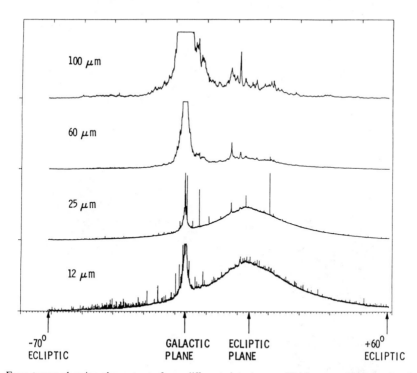

Four traces showing the outputs from different detectors as IRAS scanned across the sky. Infrared emission from the galactic plane is dominated by cool dust and so appears mainly in the 100 and 60 μm detectors. Warm dust in the solar system radiates mostly at shorter wavelengths, and so these signals reach their maximum at the ecliptic plane. Large numbers of cool stars are also apparent in the output from the 12 micron detector. (NASA).

momentum wheels unloaded when necessary by a magnetorquer system. Gas jets were not used because of the risk of contamination caused by vapours condensing on the cryogenically cooled optical surfaces. Depolyable solar panels provided 250 watts of electrical power. Since IRAS was in a low polar orbit only brief communications sessions with the ground station at the Rutherford-Appleton Laboratory near Oxford were possible, and the onboard computer was used to store instructions for an observing programme of about 12 hours. Data taken during the 12 hours of operations were stored on one of two onboard tape recorders and transmitted to the ground (at over 1 million bits per second) during the next communications session. This data was received, checked then passed to a small team of astronomers, drawn from each of the three countries, who examined the data for scientific results which might require immediate follow up. All the data was then transmitted to the NASA Jet Propulsion Laboratory in California for final analysis and catalogue production.

IRAS was launched from the Western Test Range in California on the night of 25 January 1983 (26 January in Europe). The first few days in orbit were spent carrying out an initial checkout, and apart from a malfunction in the attitude control system which generated false error signals and continuously placed IRAS into its emergency 'safe'

An impression of IRAS in orbit. (SERC).

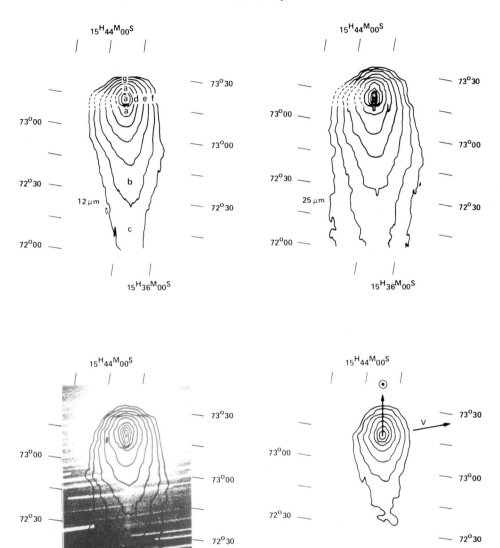

Four maps showing the infrared emission of Comet IRAS-Araki-Alcock. A comparison of the visual and infrared appearance of the comet is shown at bottom left where the IRAS contours are superimposed on a photograph of the comet. In each map the brightness at each contour is twice that of the next one outwards. The arrow on the 100 μm map (lower right) show the direction of the Sun (☉) and the direction of the comet's motion (V). A grid of celestial coordinates is marked around each map. (NASA).

mode, all the spacecraft systems were found to be in order. After intensive efforts by the control team the problem with the attitude control system was resolved by reprogramming the onboard computer, and on the sixth day the telescope aperture cover was ejected. A further few days were spent carrying out calibrations and checkout of the scientific instruments before survey observations commenced on 9 February.

IRAS operated for a total of 300 days before its liquid helium coolant was exhausted on 22 November 1983. During the mission, 95% of the sky was observed with at least two hours confirming scans, a further 3% was covered by only one hours confirmed scan, and the remaining 2% was not observed at all. In addition many thousands of additional pointed observations were carried out. The survey data were processed to produce a series of infrared catalogues, the largest of which was the IRAS point source catalogue, which was released in late 1984 and which contains about 250 000 sources. Smaller specialist catalogues released later included a list of small extended sources and a catalogue of asteroid and comet data. These catalogues represent an enormous collection of data which will be of use to astronomers for many years, and from which many important new discoveries will be made. However, the value of IRAS data was convincingly illustrated by a number of discoveries made while the mission was progressing.

John Davies and Simon Green searched the data for asteroids and comets by examining pairs of sources which were rejected from the survey because they were moving and did not 'hours confirm'. Potentially interesting sources were identified from amongst the many hundreds of spurious detections, and their positions were relayed to optical telescopes for confirmation. This effort resulted in the discovery of six new comets, including comet IRAS-Araki-Alcock which passed close to the Earth in May 1983, and several interesting asteroids including one, now named minor planet 3200 Phaethon, which may be an extinct comet. The moving object team also discovered a huge, and previously unsuspected, dust trail associated with comet Tempel-2. Similar trails associated with other comets were later found by astronomers using the more refined data products generated by the final processing in California.

Another important result was the discovery, announced by astronomer Frank Low, of a torus of warm dust in the region of the asteroid belt. IRAS data showed that the dust, which produces the zodiacal light by scattering sunlight back towards the Earth, extended 9° above and below the ecliptic plane and appears to have a banded structure. The reason for these 'zodiacal bands' is not yet fully understood, but they probably result from collisions between main belt asteroids.

More dust was discovered by H. H. (George) Aumann and Fred Gillett who found that the nearby star Vega was much brighter at infrared wavelengths than expected. After initial fears that there was a problem with the detectors (Vega was being used to calibrate the instrument at the time) the excess radiation was confirmed, and it became clear that the star is surrounded by a large, cool shell of relatively large dust grains, many times larger than normal interstellar grains. This orbiting material may be a planetary system at a very early stage in its formation, although more work is required to confirm this. A number of other stars with similar excess infrared radiation were also discovered, and these are now being intensively investigated from the ground and from airborne infrared telescopes.

An unexpected discovery was the detection of streaky patches of 60 and 100 μm emission, looking superficially like terrestrial cirrus cloud, spread all over the sky. It

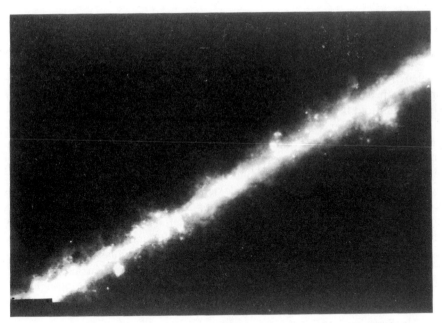

An infrared image of the plane of our Galaxy looking in the direction of the galactic centre. The knots and blobs are giant clouds of interstellar gas and dust heated by stars close to them. (NASA).

was first thought that this 'infrared cirrus' might have been caused by the combined effect of millions of distant comets, but later observations showed that it lay well beyond the Sun. The cirrus seems to originate in cool interstellar dust clouds consisting mostly of graphite grains. Some of the dust is associated with clouds of neutral hydrogen gas discovered and mapped by radio astronomers, but other patches of cirrus do not appear to match with known hydrogen clouds.

IRAS also provided much new information on the birth of stars. Observations of dense interstellar clouds revealed many infrared sources, some of which are very young stars still deeply embedded in the dust from which they formed. An important feature of these observations is that some of the infrared sources detected are probably the progenitors of low mass stars like the Sun. Although the visible light from these stars cannot penetrate the dust, they are easily detected at infrared wavelengths, and studies of them will provide new insights into the process of star formation because this class of objects has been very poorly observed in the past. Before IRAS, most studies of star forming regions had concentrated on observing more massive, and hence brighter, stars. Other sources detected in the clouds are too cool to be stars and probably represent protostellar condensations, dense regions which will one day collapse to form new stars.

IRAS also detected stars approaching the end of their lives, a period when, as a consequence of drastic changes in their internal structure, stars may lose large amounts of mass into space. Dutch astronomers showed that the red giant star Betelgeuse was surrounded by three giant rings of dust, the outermost about 1 parsec in diameter and probably ejected from the star 50–100 thousand years ago. Other dying stars, the so called OH/IR stars originally detected by radio astronomers, were found to be surrounded by dense clouds of dust. This dust apparently originates as material ejected from the ageing star which condenses as it moves away from the star and cools.

The IRAS catalogues also contain many tens of thousands of galaxies, and astronomers found, as expected, that most elliptical galaxies are faint at infrared wavelengths. This is because the elliptical galaxies are relatively dust free and have almost stopped forming stars. In contrast, normal spiral galaxies emit about half their energy in infrared because they contain numerous star forming regions which appear bright at infrared wavelengths due to of emission from warm dust.

More surprising was a class of galaxies observed by IRAS which emit enormous proportions of their total energy, in some cases well over 90%, at infrared wavelengths. This unusual brightness is often attributed to intense episodes of recent star formation, and such objects are called 'starburst galaxies'. In many cases these starburst galaxies seem to have relatively close neighbours, supporting ideas that interactions between galaxies lead to periods of enhanced star formation. It is also possible that there may be a link between the infrared bright galaxies and energetic active galaxies and quasars. It may be that some of the so called starburst galaxies are in fact quasar like objects deeply embedded in dust which absorbs the energy emitted from the central engine and re-emits it in the infrared. Only time, and a great deal more research, will tell.

6.2.3 The Spacelab-2 Infrared Telescope

IRAS was designed to detect and survey celestially fixed point sources, that is, stars and galaxies, and was only able to map diffuse emission such as the zodiacal bands and the infrared cirrus because the satellite's detectors proved much more stable than originally expected. This stability allowed adjacent scans to be added together to build up large images of the sky on which diffuse infrared emission could be traced. In contrast, the Spacelab-2 Infrared Telescope (IRT) was specifically designed to investigate emission from extended infrared sources. A secondary objective was to evaluate the suitability of the Spacelab/Space Shuttle combination as an astronomical platform and to verify the design of the instrument's cryogenic cooling system. The Spacelab-2 mission, carried in the Space Shuttle Challenger between 29 July and 6 August 1985, did not require the use of a pressurised module; all the experiments were located on pallets in the payload bay and controlled either from the Shuttle's cabin or from the ground.

The IRT consisted of a 15 cm, $f/4$, telescope operated in an off axis configuration, and a separate cryogenic tank containing 250 litres of superfluid helium. The use of a separate helium tank was possible because the short duration and large lifting capability of a Spacelab mission avoided the need to optimise the mass of the cryostat/telescope combination to ensure the maximum possible lifetime. The IRT used a suitably modified commercial helium dewar, reducing development costs considerably. The dewar was connected to the telescope by a rotary joint which allowed the telescope to scan across the sky independently of the Space Shuttle. Helium in the dewar boiled slowly away and the cold gas flowed into the telescope via a porous plug, cooling the telescope structure to about 8 K. Some of the gas was diverted to a 'cold finger' which cooled the telescope's focal plane to 3 K.

At the focal plane was an array of ten semiconductor detectors covering six wavelength bands between 2 and 120 μm. Four of these bands were fairly broad, 4.5–9.5, 8.5–14.5, 18–30, and 70–120 μm, whilst the two remaining bands were narrower,

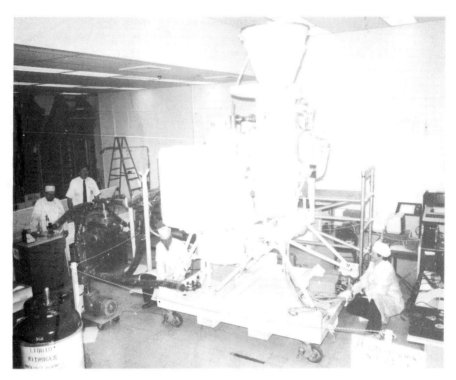

The Spacelab-2 Infrared Telescope before installation in the Space Shuttle payload bay. (NASA/Smithsonian Astrophysical Observatory).

operating in the ranges 2–3 and 6.1–7.1 μm. Only the two long wavelength bands covered the entire field of view of the telescope; the 2–3 μm detector was used to locate stars to help determine the telescope's pointing direction, whilst 6.1–7.1 and 4.5–9.5 μm bands were included to monitor water vapour in the environment around the Shuttle. The 8.5–14.5 μm detector was included because it would be sensitive to the presence of carbon dioxide released from the Shuttle.

When in use, the IRT nodded perpendicular to the Orbiter cargo bay, sweeping out a band of sky 90° wide every 15 seconds. Combining this with the motion of the Orbiter around the Earth, the experiment was able to observe about 60% of the sky per orbit, and this coverage was increased to almost 90% by rolling the orbiter 30° to either side of its nominal attitude on subsequent orbits. At times when the experiment was not in use a cooled aperture cover was placed across the telescope to protect it from contamination.

To minimise the effects of radiation hits, each detector was masked with an opaque strip down the centre. As the telescope swept across the sky, genuine infrared sources seemed to switch on, off, and on again as they were scanned, producing a characteristic signal somewhat similar to the IRAS seconds confirmation process. To eliminate detections of dust and other debris in the vicinity of the Orbiter, data from successive orbits were compared, and, as in the case of IRAS hours confirmation, detections which did not re-occur on subsequent scans were rejected.

In orbit it was found that the levels of infrared background radiation were very high, and some of the detectors became saturated, making it impossible to undertake astronomical observations with them. Some of the bright background was due to a piece of Mylar which had come loose and had lodged across the telescope aperture, but the remainder was due to infrared emission from a cloud of ice crystals and water vapour which surrounded the Orbiter. Despite these problems, which seriously degraded data from the intermediate wavelength detectors, about 12 orbits worth of usable data was obtained in the 2–3, 5–9, and 77–115 μm bands. Although the astronomical data required considerable analysis after the flight, one result was immediately obvious: the environment around the Orbiter was much more contaminated than expected, making it a very poor platform for infrared astronomy.

A second flight of the IRT was originally considered, but in view of the very high levels of infrared background found around the Shuttle and the disruption to the Spacelab programme after the explosion of Challenger in 1986, it is not clear if this will now occur.

6.3 THE OBSERVATORIES

6.3.1 The German Infrared Laboratory (GIRL)

The GIRL was to have been a cryogenically cooled 50 cm infrared telescope carried in the payload bay of the Space Shuttle and used for pointed observations of individual infrared sources. The telescope would have been mounted on the ESA instrument pointing system and equipped with a suite of four instruments including spectrometers, a photopolarimeter, and an infrared camera. Although development proceeded as far as the production of various engineering models, the project was cancelled in the mid 1980s for financial reasons.

6.3.2 The Infrared Space Observatory (ISO)

The Infrared Space Observatory was proposed to ESA in 1979 and selected for detailed study (Phase A) in March 1983. Phase B studies, lead by the French company Aeropatiale, began in December 1986. The design closely resembles that of IRAS, consisting of a cooled 60 cm telescope mounted on a separate service module. The service module contains the power and communications systems together with momentum wheels used to point and stabilise ISO with an accuracy of a few arc seconds. The payload module is a large dewar containing 2040 litres of liquid helium surrounding the telescope and its scientific instruments. Like IRAS, the ISO telescope will be protected by an ejectable aperture cover before launch and for the first few days in orbit. Panels fitted on the Sun facing side of the satellite act both as a sunshade and support the solar cells. ISO will be approximatly 5.2 m tall and will weigh 2.3 tonnes at launch.

ISO is designed to study individual infrared sources in detail, and will carry four scientific instruments, each developed by separate international consortia. The four instruments all have higher spectral and spatial resolution than the IRAS satellite and are described briefly below. One of the factors which enables the ISO instruments to improve on the capability of IRAS has been the development of infrared detector arrays, matrices of tiny semiconductor detectors mounted on a single electronic chip.

The observatories

Infrared arrays can provide much better spatial resolution than the single element detectors used in earlier instruments, making it possible to develop high quality infrared cameras, and can also be incorporated into spectrometers.

The ISO infrared camera (ISOCAM) will operate between 2.5 and 17 μm and will be able to image a field of view up to 3 arc minutes across. Light entering the camera will be sent by a mirror to one of two channels, each equipped with a 32 × 32 element infrared array detectors. One channel will cover the 2.5–5.5 μm band; the other will operate between 5 and 17 μm. Wheels in each channel will be used to select one of four lenses to provide image scales varying from 1.5–12 arc seconds per pixel, and to position various

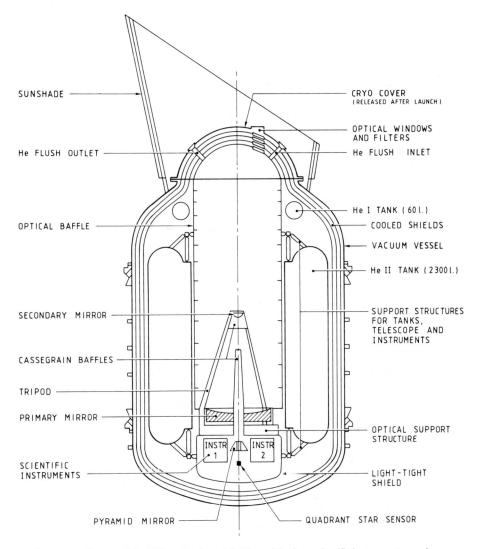

A cutaway diagram of the ISO payload module. Two of the four scientific instruments are shown beneath the main mirror. Note the relatively small size of the telescope compared with the entire payload module which stands about 3.5 metres tall (excluding the sunshade). (ESA).

narrow band filters in the beam. Circular variable filters, sheets of material coated in such a way that different zones transmit different wavelengths, will also be fitted. By scanning a circular variable filter in front of a detector and measuring the flux at regular intervals it is possible to produce a series of images in different wavelength bands. The ISO circular variable filters will have a resolution (λ/deltaλ) of about 50.

The photopolarimeter (ISOPHOT) will consist of four subsystems each optimised for different types of observations. These include multiband, multiaperture photometry from 3–110 μm, a photometric camera mode covering the 30–200 μm range, a system of linear arrays which can be used to map extended sources in the 3–30 μm range, and two spectrophotometers operating from 2.5–12 μm. Different combinations of the four subsystems can be chosen by the appropriate setting of selector, aperture, and filter wheels.

The short wavelength spectrometer will operate between 2.3 and 45 μm, a region of the spectrum which contains many important infrared spectral lines. It is split into two almost independent sections. Light entering the instrument is resolved as it is reflected from one of two diffraction gratings and then directed, via a series of slits, filters, and mirrors, onto small arrays of detectors, each optimised for different wavelength regions. The spectral resolution of the instrument will normally be about 1000, but in the 15–35 μm range this can be increased to about 20 000 by the use of a device known as a Fabry–Perot interferometer.

The long wavelength spectrometer will cover the wavelength range from 45–180 μm with a low resolution mode of about 200. It uses a diffraction grating to resolve the incoming beam and direct it, via a refocusing mirror, onto an array of ten detectors. To obtain a detailed spectrum, the grating is then rotated in steps through a few degrees and the output of each detector is read after each movement of the grating. After a number of such steps, the range of wavelengths covered by each detector will just overlap with that covered by its neighbours, and a complete spectrum can be constructed by computer processing of the data from each detector. To achieve higher spectral resolution, typically λ/delta λ of 10 000, a Fabry–Perot interferometer will be placed into the optical path.

The four instruments will be mounted in 90° segments below the mirror of the ISO telescope and will view the sky via a pyramidal mirror mounted close to the Cassegrain focus of the telescope. Each instrument will view a slightly different region of sky, and the one required will be selected by adjusting the pointing of the telescope slightly. Normally, only one instrument will be operated at a time, but it may be possible to use a second instrument, probably the camera, in a parallel mode.

ISO is expected to be launched by an Ariane 4 rocket in about 1993 and to be placed in a highly elliptical 24 hour orbit of low inclination. In this orbit the satellite will spend about 14 hours in continuous contact with a single ground station as it climbs to its 70 000 km apogee and then sweeps back towards perigee. This will allow ISO to be operated as an observatory with astronomers at the ground station able to control the satellite throughout their observations. The highly elliptical orbit also allows long, uninterupted observations of individual sources to be made, a huge improvement on the 8 minute observations possible with IRAS. Scientific data are not stored aboard the spacecraft, but are transmitted to the ground station as they are collected.

If only one ground station is available, then contact with ISO will be lost for the 10 hours around the time of perigee. However, since the satellite spends some of this time in the Earth's radiation belts, where charged particle hits render the detectors inoperative, only about 8 hours of observing time will be lost per orbit. If a second ground station is available, possibly as a result of international cooperation between ESA and another agency, this additional observing time could be recovered. The decision to place ISO in a 24 hour orbit was taken in May 1987 after studies showed that the reduced radiation background in the higher orbit would provide more good quality data than the 12 hour orbit originally planned could obtain using two ground stations.

6.3.3 The Space Infrared Telescope Facility (SIRTF)

SIRTF grew out of a 1974 recommendation from the American National Academy of Sciences that NASA should develop a cooled infrared space telescope 'as soon as the results of the infrared sky survey satellite have been digested'. The SIRTF was conceived as a Spacelab payload carried in the Space Shuttle cargo bay for missions of 7 to 30 days. This concept envisaged a liquid helium cooled, 85 cm telescope mounted on the Spacelab instrument pointing system. A suite of up to six scientific instruments were to have been carried, with full operating capability being reached after several missions with a smaller scientific payload had proved the basic design of the telescope.

By 1984, delays and rising costs associated with the Space Shuttle programme led to a decision to redesign SIRTF as a free flying satellite. SIRTF became a facility class mission like the Hubble Space Telescope, requiring in-orbit servicing to achieve a long operational lifetime. Although development of SIRTF was slowed by the financial and technical crises which affected NASA in the mid 1980s, studies are continuing to refine the design concepts of the mission and to prepare for full scale development when funds become available.

At present SIRTF is envisaged as a cooled $f/1.5$ telescope with an 85 cm clear aperture and a pointing accuracy of 0.25 arc seconds. The liquid helium supply is expected to last about two years, requiring at least two servicing missions to meet the minimum six year lifetime expected. The satellite will be equipped with a complement of instruments making it a versatile space observatory, and the initial payload is expected to include an infrared camera operating between 3 and 30 μm, an imaging photometer covering the 3–700 μm range, and a 2.5–200 μm spectrograph.

Various orbits have been considered for SIRTF including a polar orbit at about 900 km and a low, 3–400 km, orbit inclined at 28°. The high polar orbit provides better observing conditions and reduced risk of contamination from the traces of atmospheric gases found in low orbit, but the low 28° orbit is accessible from the proposed international space station, making in-orbit servicing easier. Depending on the orbit chosen, SIRTF may be based around a specially built spacecraft (possibly developed jointly with the AXAF programme), carried on a Multi Mission Spacecraft or fitted to a co-orbiting platform developed as part of the space station.

6.4 STUDYING THE COSMIC BACKGROUND RADIATION

6.4.1 Introduction

The cosmic background radiation, weak but uniform emission from the entire sky, was discovered by Arno Penzias and Robert Wilson of the Bell Telescope Laboratories in 1964. Penzias and Wilson were not astronomers; they were using a large horn antenna as part of technological experiments connected with the Telstar communications satellite. Even after identifying and eliminating sources of terrestrial interference, the two men found that wherever they turned their antenna there remained a background of unexplained microwave radiation.

The existence of such a background had been predicted in the 1940s (and again in the early 1960s) as a direct consequence of 'Big Bang' cosmological models in which the Universe is expanding from its creation as a single point of infinitely high density and temperature. In the 15 thousand million years since the Big Bang, the Universe has expanded and cooled, causing its temperature to fall to just a few degrees above absolute zero. The microwave background detected by Penzias and Wilson comes from this original explosion; it is a thermal echo of the Big Bang. Other astronomers soon detected the cosmic background radiation, and measurements at far infrared, millimetre, and radio wavelengths showed that its spectrum is equivalent to that from a 2.7 K black body, a temperature consistent with the predictions of Big Bang models.

This book will not attempt to summarise modern cosmological theory, but will simply note that while in the early Universe there was a constant interchange between energy in the form of photons and energy in the form of particles, the photons became decoupled from matter about 500 000 years after the Big Bang. Since then matter has gone on to form galaxies, stars, planets, and asteroids, while the background radiation has been steadily shifted to longer wavelengths by the expansion and consequent cooling of the Universe. The background radiation is an astronomical fossil, recording conditions in the very early Universe, and may provide clues to much that has happened since the Universe was very much smaller and hotter than it is today.

The cosmic background radiation thus offers a chance to look back into the very distant past and make observations which may be crucial to cosmological theory. It is particularly important to measure the degree to which the background is isotropic (equally bright in all directions) and to search for small fluctuations which might mark density enhancements in the very early Universe from which clusters of galaxies eventually formed. Another crucial measurement is to determine the spectrum of the radiation to see how closely it matches a black body curve. Failure to follow the predicted spectrum may indicate the existence of additional energy sources in the early Universe, for example radiation emitted from material falling into massive black holes. It may even be possible to detect the faint glow from the first generation of galaxies superimposed on the emission from nearer, and much brighter, material such as interplanetary and interstellar dust.

Space experiments offer certain advantages for such studies, notably the opportunity to observe at wavelengths impossible from the ground and the chance escape from the effects of varying transparency in those atmospheric windows which are available. Space observations also enable the entire sky to be studied by a single instrument, avoiding he need to calibrate different telescopes situated in different hemispheres. Some such observations have already been attempted, and others are planned.

6.4.2 Relikt-1 and 2

The Soviet Prognoz-9 satellite, launched on 1 July 1983 into a highly elliptical orbit, carried a number of solar experiments, X-ray and gamma ray burst instrumentation and a pair of small radiotelescopes known as RELIKT-1, an experiment to search for any

A Soviet Prognoz satellite of the type which carried the RELIKT-1 experiment. (CNES).

anisotropy (unevenness) in the cosmic background radiation. The satellite was placed in a highly elliptical orbit with a period of 27 days and rotated about its Sun vector every 2 minutes. One of the RELIKT antennae pointed at right angles to the Sun vector and swept out a great circle on the sky every revolution; the second antenna pointed directly away from the Sun. The RELIKT instrument, a Dicke radiometer passively cooled to about 100 K, continually compared the signals from the two antennae to search for any variation in the background radiation. The satellite was re-oriented once a week until the radiotelescope, operating at a wavelength of 8 mm, had surveyed the entire sky. This process took 6 months to complete. The experiment placed upper limits on the anisotropy of the background, a result which constrains theoretical models of the early Universe, and showed that it would be possible to design an instrument sensitive enough to distinguish between a variety of cosmological models.

A proposed follow up experiment, RELIKT-2, will be mounted on a spacecraft in orbit around the gravitationally stable L2 liberation point, about 1.5 million kilometres from the Earth and opposite to the Sun. This unusual orbit has been chosen to minimise the effects of thermal emission from the Earth and the Moon. The RELIKT-2 instrument will operate in five different wavelength bands between about 1.5 mm and 15 mm. Like RELIKT-1, the detectors will be radiatively cooled to about 100 K.

6.4.3 The Cosmic Background Explorer (COBE)

COBE, a satellite designed specifically for cosmological studies, was the result of three proposals submitted to NASA in 1974. These led to the selection of a team to conduct a study in conjunction with the Goddard Spaceflight Center, and the COBE mission evolved from this work. The satellite will be entirely concerned with diffuse emission at far infrared and millimetre wavelength and will have no capability to resolve point sources.

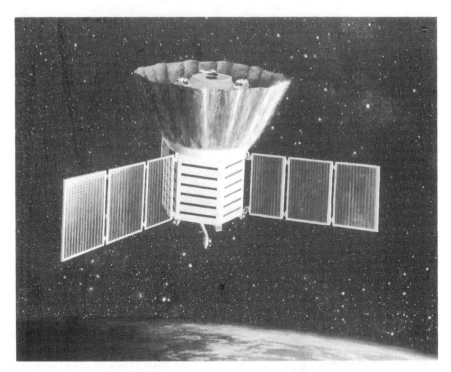

An impression of the COBE satellite following the changes required for launch on a Delta rocket. (NASA).

COBE will consist of a service module and a payload module which will carry three instruments, each of which is designed to investigate a different aspect of the background radiation. To reduce the effects of thermal radiation produced in the instruments themselves, two of the COBE experiments will be surrounded by a dewar containing about 600 litres of superfluid helium which will cool the instruments to about 2 K. The supply of coolant will determine the duration of the mission, expected to be about one year.

The Diffuse Infrared Background Experiment (DIRBE) will operate in ten wavelength bands from $1-300\,\mu$m and is designed to search for the cumulative emission from primordial galaxies formed soon after the Big Bang. The instrument, which is contained in the helium dewar, uses a 20 cm, off-axis Gregorian telescope to observe a $0.7°$ square field of view $30°$ away from the spin axis of the satellite. Radiation from the sky follows a complex optical path designed to suppress stray light and passes through various filters and polarisers until it reaches an array of detectors comprising indium antimonide photoconductors, doped silicon and germanium photodiodes, and bolometers. A vibrating beam interrupter, operating 32 times a second, causes the detectors to switch between measuring radiation from the sky and observing a 2 K cold stop within the instrument. Comparing the signal from the sky to the signal measured when observing the cold stop will allow the DIRBE to operate as an absolute radiometer and produce an accurate map of diffuse infrared emission. Known sources of radiation, such as starlight and emission from warm dust, will be subtracted from the maps (using polarisation information to help distinguish this foreground radiation from the cosmic background) and, once all the known sources have been removed, any signal which remains will be due to the cosmic background.

Also located in the helium cryostat in the Far Infrared Absolute Spectrophotometer (FIRAS), an instrument designed to determine the spectrum of the cosmic background radiation. The FIRAS, which looks along the spin axis of the satellite, has a 7° field of view and operates from 100 μm to 10 mm with a resolution of 5%. The seven degree field of view means that the FIRAS will record the spectrum of the background radiation at approximately 1000 different points on the sky during the COBE mission.

The FIRAS is basically a Michelson interferometer, a device which splits an incoming beam of light into two equal parts, delays one of them slightly relative to the other, and then recombines them before sending them to a detector. The two parts of the beam will recombine perfectly if the delay corresponds to an exact number of wavelengths, or will interfere with each other, cancelling themselves out, if the delay corresponds to an odd number of half wavelengths. By moving mirrors within the instrument the delay before recombination can be changed, and the result is a varying signal, known as an interferogram, recorded by the detector. The interferogram contains information about the spectrum of the incoming radiation, and this can be extracted mathematically.

The FIRAS uses a trumpet shaped cone to collect radiation from the sky and send it to the interferometer. The temperature of the cone can be controlled electrically so that the emission from the cone can be measured and allowed for. A black body radiator, the temperature of which can be controlled with an accuracy of 0.001 K, can be inserted into the mouth of the cone, blocking off the sky and providing a signal of precisely known intensity for calibration purposes. The FIRAS is equipped with four bolometer detectors and is expected to produce a spectrum of the background radiation accurate to 0.1%.

The Differential Microwave Radiometer (DMR) is designed to determine the isotropy of the background at wavelengths of 3.3, 5.7, and 9.6 mm. It is based on similar instruments flown on balloons and aircraft and uses a microwave receiver to monitor the total amount of power detected at a particular wavelength. Since the weak signal from the sky may be swamped by noise produced in the receiver, the DMR switches between two antennae and measures any difference between signals from two widely separated areas of sky. By scanning the instrument rapidly (using the rotation of the satellite) effects introduced by differences between the antennae can be eliminated, and the DMR can detect any anisotropy in the background radiation itself.

The DMR has three receiver boxes, one for each wavelength, and each contains two independent receivers to improve sensitivity and increase reliability. Each box is fed by two separate antennae, each with a 7° field of view, which point 60° apart on the sky and are angled at 30° from the spin axis of the spacecraft. The receivers will be maintained at their operating temperatures, 300 K for the long wavelength channel, 140 K for the others, by a combination of thermostats, radiators, and heaters. A combination of precise thermal control, regular calibrations, and sophisticated data analysis will allow the DMR to make sky brightness maps accurate to 0.000 15 K at 5.7 and 3.3 mm, 0.0003 K at 9.6 mm.

COBE is designed to operate from a 900 km, Sun synchronous, near polar orbit similar to that used by IRAS. This orbit was chosen to allow COBE to scan the entire sky twice during its one year lifetime. The satellite will be pointed 94° from the Sun vector and rotated at 0.8 rpm to allow the DIRBE and DMR to sweep across the sky. It was originally intended to launch COBE in 1988, using a Space Shuttle to carry the satellite into a low parking orbit and an onboard rocket to boost COBE to its final altitude, but in January 1987 NASA announced that the mission would be transferred to an expendable Delta rocket and launched in 1989. The change of launch vehicle has meant reducing the diameter of the satellite from 4.5 to 2.5 metres and modifying the structure, solar array, and DMR receivers. Since the Delta launcher can lift COBE direct into its operating orbit, the onboard rocket will no longer be needed and so will be deleted.

6.5 MILLIMETRE AND SUBMILLIMETRE ASTRONOMY

6.5.1 Salyut 6 BST-1M

SIRTF and COBE will both operate at wavelengths which overlap what are traditionally regarded as the infrared and submillimetre regions, and this is also true of other instruments, broadly described as submillimetre or millimetre telescopes. Such an instrument was the BST-1M telescope fitted to the Soviet Salyut 6 space station. The BST-1M was a 650 kg instrument operating from 50 μm to 1 mm which was located in the main volume of the space station. The telescope was 1.5 m in diameter and used helium cooled semiconductor detectors. Unlike most such spaceborne instruments, the BST-1M was used for both astronomical and atmospheric studies, and for the latter purpose it included an ultraviolet detector. The instrument was first used in 1977 by the crew brought to Salyut 6 in the Soyuz 26 spacecraft and was operated at intervals throughout subsequent operations on Salyut 6. Use of the telescope was limited because a compressor used as part of the cooling system demanded almost the entire power output of the space station's solar panels.

6.5.2 Aelita

Aelita is a Soviet mission which will use a cooled telescope for photometry in the range 150 μm to 2 mm. It will use an $f/1.4$, one metre telescope with aluminium alloy mirrors. The telescope will be carried in a dewar of liquid neon which will maintain it at a temperature of 27 K. An ejectable cover, which will also contain a tank of liquid neon, will protect the telescope before launch and during its first few days in orbit. The germanium based bolometric detectors in the focal plane will be cooled to about 0.32 K by a closed cycle absorption refrigerator using helium 3, a rare isotope of normal helium (helium 4). In this system liquid helium 3 is evaporated to cool the focal plane and the vapour is adsorbed onto the surface of a palladium coated silica gel adsorbent. At regular intervals the silica gel is warmed to drive off helium 3 which is then recondensed to begin the cycle again. Energy removed from the helium 3 when it is recondensed will be transferred into a tank of superfluid helium 4 which will slowly boil away.

Aelita will carry about 400 kg of cryogens, giving a mission lifetime of 1–1.5 years. The satellite is expected to have a pointing accuracy of about 10–15 arc seconds using CCD star trackers provided by Karl Ziess, Jena. Aelita is unlikely to be launched before the middle of the 1990s.

6.5.3 The Far Infrared Space Telescope (FIRST)

FIRST is a proposed ESA mission using an 8 metre deployable antenna for observations in the range 100 μm to 1 mm. If developed, FIRST will provide unparalleled angular resolution and sensitivity in the far infrared. The spacecraft would probably be similar to ISO with a cooled payload module (but not a cooled telescope) and a separate service module.

6.5.4 The Large Deployable Reflector (LDR)

The LDR is a NASA concept to construct a large, 10–30 metre diameter, orbiting telescope for studies in the 2 μm to 1 mm range. The LDR would be launched by the Space Shuttle and deployed in orbit. The operating lifetime of the LDR would be at least 10 years, and to achieve this will require regular servicing from either the Space Shuttle or the Space Station. The LDR is unlikely to be flown until the next century.

7

Radio astronomy from space

7.1 INTRODUCTION

Radio waves from beyond the Earth were first recognised in 1932 by Karl Jansky, a physicist at the Bell Telephone Laboratory at Holmdel, New Jersey, who was carrying out research on atmospheric static, the background hiss which affects long distance radio communications. During his research Jansky discovered radio emission from the plane of the Galaxy, determined that the emission was strongest from the direction of the galactic centre, and correctly suggested that much of the radiation arose in the interstellar medium. Surprisingly, the discovery aroused little interest at the time, and it was left to an amateur astronomer named Grote Reber to follow up Jansky's work. Reber built the world's first radio telescope in the garden of his house in Wheaton, Illinois, and explored the radio sky as a hobby. Only after the Second World War, with its legacy of radio and radar research, did radio astronomy blossom into an important area of science. The pioneering days of radio astronomy, and many of the results which it has achieved, are described elsewhere and will not be recounted here.

Radio astronomy is possible because the atmosphere is transparent to wavelengths between a few millimetres and a few tens of metres, a range which corresponds to frequencies of about 30 gigahertz (GHz) to a few megahertz (MHz). Beyond this atmospheric window, extraterrestrial radio signals cannot reach the ground because the ionosphere, a region of ionised atoms and molecules extending from about 90–500 km above the Earth, reflects them. The reflecting properties of the ionosphere make it possible to bounce radio transmissions beyond the horizon and thus communicate over great distances, but also close off a potentially interesting region of the electromagnetic spectrum from Earthbound radio telescopes. Although the condition of the ionosphere, and hence the precise wavelengths which it will transmit, varies with solar activity, there are certain regions of the radio spectrum which can be observed only from space.

The first long wavelength measurements of the cosmic radio background were made by Canadian scientists in 1960, using the American Transit II navigation satellite. Other studies were made from sounding rockets and from satellites in the Aloutte, Ariel, and Electron series, although these satellites were more concerned with studies of the ionosphere and the Earth's radiation belts than with extraterrestrial radio emission. Other experiments were carried on satellites designed to travel well beyond the outer fringes of the ionosphere on lunar (Zond 2 and 3, Luna 10 and 11) or interplanetary (Venera 2) missions. These simple experiments had little or no directional sensitivity but did detect extraterrestrial emissions in the wavelength range from 30–1500 m.

Probably the most important results from these early experiments were measurements of the spectrum of the radio background which showed that the background reached its maximum intensity at a wavelength in the region of 30–300 m (corresponding to frequencies between 10 and 1 MHz) and then fell away owing to free–free absorption of radio waves by interstellar electrons. Also of interest was the detection of radio bursts occurring high in the Sun's corona. These experiments also identified various problems with both instruments and operating techniques which needed to be solved before advanced radio astronomy satellites could be developed. Amongst the changes required were improvements in the long term stability of the radio receivers, a reduction in the amount of radio noise developed by the satellites themselves, and the need to place future radio astronomy missions into very high orbits. This latter requirement stems from the necessity to reduce the effects of the ionosphere which can act as a source of radio noise and as a refracting medium for radio waves, and may also subtly alter the properties of a spacecraft's radio antennae. So significant are the effects of the ionosphere that in 1964 Francis Graham Smith cautioned that 'low frequency radio astronomy from satellites is always in danger of becoming a geophysical rather than an astronomical enterprise'.

7.2 SECOND GENERATION RADIO ASTRONOMY EXPERIMENTS

Radio astronomy satellites are listed in Table 7.1

7.2.1 Radio Astronomy Explorer-1

The first satellite with appreciable angular resolution at radio wavelengths was Radio Astronomy Explorer (RAE) 1, otherwise known as Explorer 38. The RAE-1 mission (one of a pair of satellites) grew out of studies at the Goddard Spaceflight Center in 1963, and detailed development began in 1964. The stated objectives of the RAE programme were:

(1) To determine the spatial and spectral distribution of cosmic radio noise.
(2) To measure solar and Jovian radio bursts.
(3) To detect discrete cosmic radio sources.

Construction and testing of the spacecraft, a one metre tall cylindrical body with four solar panels on extending paddles, took place in 1966 and 1967.

The satellite, which weighed about 200 kg, was launched by a Thrust Augmented Improved Delta rocket from the Western Test Range in California on 4 July 1968. After insertion into an elliptical transfer orbit, the satellite used a Thiokol TE-M-479 solid rocket motor to move into a circular orbit at an altitude of 5800 km and an inclination of 121°. The orbit was a compromise between the need to rise above the outer fringes of

Table 7.1
Radio astronomy satellites

Name	Launch date	Launch vehicle	Perigee (km)	Orbit† Apogee (km)	Inclination(°)	Notes
RAE-1 (Explorer 38)	4 Jul 1968	TAID‡	5829	5864	120.9	
Explorer 43	13 Mar 1971	Long tank Thor-Delta	146	122 146	28.7	Radio astronomy experiment on IMP spacecraft
RAE-2 (Explorer 49)	10 Jun 1973	Delta 2913	Lunar orbit,	100 km circular	59	Last US lunar Mission
Salyut-6 KRT-10	28 Jun 1979	A-2	395	405	51.6	Launch and orbital data for progress 7 supply craft which delivered KRT-10 to Salyut-6

†Since satellite orbits change because of atmospherc drag etc., orbital parameters quoted by different sources may vary
‡Thrust Augmented Improved Delta

the ionosphere, a desire to remain close enough to the Earth so that gravitational effects could stabilise the spacecraft, the power of the available launch vehicles, a requirement that the satellite be in continuous sunlight for as long as possible to minimise the effect of changing thermal conditions on the radio antenna, and the need to scan as much of the sky as possible in the planned one year mission.

Once in its final orbit the satellite's spin rate (about 80 rpm during the solid rocket motor burn) was reduced to zero by using a magnetorquer, and RAE-1 began a slow metamorphosis as long, thin radio antennae began to grow slowly outwards. The antennae were composed of flat metal tapes, 0.005 cm thick, which were launched wrapped around wheels within the spacecraft. The tapes automatically curled to form long, 1.3 cm diameter, tubes as they were unwound from the satellite, and interlocking tabs on the edges of the tape meshed during the deployment process to keep the antennae rigid. The deployment was carried out in a number of stages and monitored by TV cameras on the satellite which viewed targets on the ends of the antennae. Deployment was complete by October 1968 by which time RAE-1 boasted an X shaped array composed of four antenna each 229 m long, a dipole antenna composed of two elements 19 m long, and a long boom, over 200 m long, used to damp unwanted motions of the satellite. To avoid distortions caused by uneven thermal expansion, the outsides of the booms were coated with silver and the insides painted black. Holes in the tape allowed some sunlight to reach the inside of the booms and warm them, reducing the temperature gradients along and across the antennae.

RAE-1 was stabilised so that one half of the X-shaped array pointed towards the Earth and the other looked outwards towards space. It was maintained in this attitude by the differential pull of gravity on the lower part of the array compared with the upper part, a situation known as gravity gradient stabilisation. In this condition RAE-1's upper antennae swept out a swath of sky during every 3.75 hour long orbit with the TV cameras being used to determine the attitude of the spacecraft and to identify the regions of sky from which radio signals were being received. The lower portion of the array served to monitor radio emission from the Earth, both as a scientific study and for comparison with signals detected by the upper antennae.

The RAE carried two types of radio receiver, one type used for mapping of the radio background, the other for monitoring radio bursts such as those emitted by the Sun. The mapping instruments were Ryle–Vonberg radiometers able to record radiation at nine different wavelengths from about 1500 m to 33 m (frequencies of 0.2 to 9.18 MHz). This type of instrument was chosen because its stability made it suitable for mapping large areas of sky over long periods. Two such radiometers were connected to the upper, sky scanning, V, with one acting as a reserve for the other to ensure a long mission lifetime. Only one Ryle–Vonberg radiometer was connected to the lower, Earth pointing, V. Examples of the second type of instrument, known as a burst radiometer, were connected to the lower V and the short dipole antenna. The burst radiometers could record data at a rate of 2 samples per second, much faster than the Ryle–Vonberg radiometers, and were used to monitor transient phenomena such as solar radio bursts. The burst radiometer connected to the dipole antenna also had a channel which was scanned through 32 wavelength steps every 8 seconds to record the radio spectrum of rapidly changing sources. The data obtained were either transmitted to ground stations immediately or recorded on board for later playback.

The huge X shaped array provided an angular resolution of about 20° for observations at a wavelength of 150 m and the combination of the satellite's orbital motion, and the precession of the orbit around the Earth (at a rate of 0.5° per day) allowed all the sky between declinations of 60° and −60°, limits set by the orbital inclination of the satellite, to be mapped in just under one year. The resulting maps, although plagued by various forms of interference, showed that the majority of long wavelength radio waves, which are probably caused by fast moving electrons being curved by interstellar magnetic fields, arise within our galaxy. It was also found that the intensity falls dramatically in some regions of the Galactic plane owing to absorption of the radio signals by interstellar gas. The wavelength at which the emission from our Galaxy peaked was found to be about 100 m with a steady reduction at longer wavelengths due to free–free absorption by electrons. RAE-1 also detected signals at longer wavelengths, reaching a maximum at about 1200 m, which seem to originate beyond our Galaxy.

RAE-1 also observed sources within our solar system, recording large numbers of radio bursts from the Sun during the mission. Attempts to detect radio emission from Jupiter were initially unsuccessful, but later analysis did show radio bursts from the giant planet in the wavelength range from about 60–660 m (4700 kHz–450 kHz). Radio noise originating in terrestrial thunderstorms and high in the Earth's atmosphere was also identified in RAE-1 data, and, although providing a fruitful area of study for some workers, this was a major problem for astronomers interested in galactic mapping. In some cases observations from RAE-1 were coordinated with observations at shorter wavelengths made with ground based telescopes. Measurements were also coordinated with rocket experiments to provide an independent check of the RAE-1 results. One such launch was on 15 October 1969 when an Astrobee sounding rocket was launched to an altitude of 2576 km from the NASA facility at Wallops Island as RAE-1 passed overhead.

Despite various problems, including the failure of the spacecraft's tape recorder after only two months and a series of malfunctions within the scientific instruments, RAE-1 operated for a total of four years. Overall, the mission was successful, especially in terms of solar, Jovian, and terrestrial studies, but no catalogues of extra-solar radio sources were ever published.

7.2.2 Explorer 43

Explorer 43 was not a dedicated radio astronomy mission but was one of a series of satellites known as Interplanetary Monitoring Platforms (IMP). The IMP spacecraft, which were all part of the Explorer series, were placed in highly elliptical orbits from which they could monitor the solar wind and interplanetary magnetic fields beyond the immediate vicinity of the Earth. Explorer 43 was drum shaped, about 2 m high and 1.5 m in diameter, and was spin stabilised at a rate of 5.4 rpm. As well as instruments for the study of cosmic rays, energetic particles, plasmas, and magnetic fields, the spacecraft carried a radio astronomy experiment similar to that on RAE-1. Explorer 43 was launched on 13 March 1971 into a highly eccentric Earth orbit.

The radio astronomy instrument consisted of a set of deployable antennae which formed a pair of dipoles about 100 m long. The dipoles were connected to two radio receivers each able to determine the spectrum of the radio background in 32 separate channels in the wavelength range from 30–10 000 m (10 MHz–30 kHz). Explorer 43 detected radio emissions from the Sun and the Earth as well as successfully measuring the spectrum of the galactic radio background between 110 and 2300 m (2.6 MHz–130 kHz). The results were in generally good agreement with data from RAE-1, and extended to longer wavelengths (lower frequencies) than the results from the earlier mission. Explorer 43 re-entered the atmosphere on 2 October 1974.

7.2.3 Radio Astronomy Explorer-2

The discovery that natural and man-made emission from the Earth were both common and powerful at the wavelengths to which the first Radio Astronomy Explorer was sensitive led to the second spacecraft being placed in orbit around the Moon. The reasons for this choice were that in lunar orbit terrestrial interference would be generally much reduced and for some of the time RAE-2 would be behind the Moon and thus completely screened from the Earth's radio noise. A further advantage was that since the Moon has no ionosphere, the satellite could study a greater range of frequencies (25 kHz–13 MHz) than its predecessor.

RAE-2, also known as Explorer 49, was similar to RAE-1 except for changes required to enable it to operate in lunar orbit and improvements in the scientific instruments. The most obvious differences were the addition of a solid propellant rocket motor, used for lunar orbit insertion, on the top of the spacecraft, and a separate hydrazine system, for velocity control during the trip to the Moon, on the bottom. RAE-2 was spin stabilised during its translunar coast and three axis stabilised a few weeks after reaching lunar orbit. Attitude control was provided by a combination of freon gas thrusters and gravity gradient techniques.

Second generation radio astronomy experiments

The Radio Astronomy Explorer 2 is prepared for launch. (NASA).

The 328 kg spacecraft (200 kg plus boost motors) was launched by a Delta 2913 rocket on 10 June 1973 and was inserted into a circular orbit 1100 km above the Moon on 15 June. The orbit was inclined at 59° to the lunar equator and precessed at a rate of 0.14° per day, so that in one year the band of sky scanned by the satellite moved about one sixth of the way around the celestial sphere. Each orbit lasted 222 minutes, and, at certain times of the month, up to 48 minutes per orbit were spent behind the Moon. During these periods data were recorded and then transmitted to mission controllers when the satellite was in view from the ground. After lunar orbit insertion RAE-2 deployed its 38 m dipole antenna and operated in a spin stabilised mode for several weeks. After this period the dipole antenna was stowed, the spacecraft was re-oriented and stabilised, the long V antennae and liberation damper were extended, and the dipole antenna redeployed. Unfortunately, part of the lower V antenna did not extend correctly and finished up parallel to the local vertical, leaving the lower V in an asymmetric condition. Because of this problem the lower V was extended to only 183 m during the first 16 months of the mission, with deployment to the full 229 m length being delayed until November 1974.

The location of RAE-2 in lunar orbit allowed it to observe the Earth in the same way as it monitored other celestial bodies, and during the mission terrestrial radio emissions were detected from both thunderstorms and from electrons in the auroral zones. Separating these signals from those caused by other celestial sources was greatly assisted by the regular periodicities caused by the orbital motion of the satellite around the Moon and of the Moon around the Earth. Terrestrial signals were also cut off at regular intervals as RAE-2 moved behind the Moon. Other radio sources within the

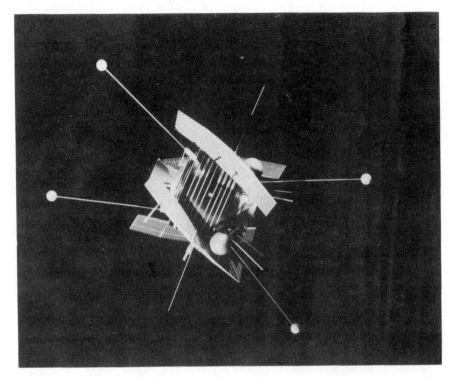

Model of the Radio Astronomy Explorer 2 satellite in orbit. Note that the antennae are only partly deployed. (NASA).

solar system, notably the Sun and Jupiter, were also observed. Measurements of the radio emission from the Galaxy were complicated by the emission from the Sun, the Earth, and Jupiter, and careful processing of data taken over almost three years of observations was required to make reasonably error free maps of galactic radio emission.

A detailed discussion of the interpretation of RAE-2 data would be out of place here, since the results do not concern point sources of radio emission but deal with large and diffuse structures within the interstellar medium which were observed with an instrument which had very poor angular resolution. At very long wavelengths (that is, at low frequencies) the radio waves are absorbed by electrons within a few parsecs of the Sun, and at these wavelengths the sky appears more or less uniform. At shorter wavelengths radio waves can travel further before being absorbed, and it is possible to detect radiation which arises towards the centre of our Galaxy and even from extragalactic space. Little galactic radio emission is observed in some directions, and this may be due to regions where the interstellar magnetic fields are weak, where fast moving electrons are few, or to areas where the diffuse radio emission from distant regions is absorbed by particularly high concentrations of ionised material in the foreground. In other directions there are zones of increased radio emission which appear to arise in curved structures; these may be the remnants of very old supernovae. Several such remnants are expected to lie within a few hundred parsecs of the Sun and could well be detected by the instruments on RAE-2.

Much further work will be needed to fully understand the structure of the interstellar medium, and it may be that in the future data from the Radio Astronomy Explorers will be combined with soft X-ray maps and data from the Extreme Ultraviolet surveys described in section 4.12 to map out the material between the stars in greater detail.

7.2.4 Salyut 6 KRT-10 (Cosmic Radio Telescope-10)

This 10 m antenna was carried to the Soviet Salyut 6 space station by the unmanned Progress 7 supply craft. The 300 kg instrument, which consisted of a deployable antenna attached to a long pole plus various electronics boxes and power leads, was first assembled in the main work area of the space station. It was then fitted to the rear docking port of Salyut 6 in such a way that it was outside the pressurised volume of the Salyut but extended into the cabin of the still attached Progress spacecraft. After the Progress supply ship had been undocked, the telescope was left projecting into space and was then unfurled remotely by the crew inside Salyut 6.

The KRT-10's observations, which included studies of the radio emission from the Sun, the Galaxy, and pulsar PL0329, were overshadowed by the drama which ensued when the cosmonauts attempted to jettison the antenna after three weeks of observations. Instead of drifting away from Salyut 6, the KRT-10 became entangled on projections around the edge of the docking port. Although the crew could still return to Earth by using the Soyuz craft docked at the opposite end of Salyut 6, the blocked rear port prevented further supply ships from docking with the station. Since this would have severely curtailed the mission the two cosmonauts donned space suits, spacewalked to the antenna, and jettisoned it manually.

7.3 SPACE VLBI

The long wavelengths of radio waves compared with the wavelength of visible light means that the ability of a single radio telescope to see fine detail, that is, its resolving power, is very poor[†]. The resolving power of radio telescopes can, however, be improved by observing a source with two or more telescopes and combining the signals from each instrument. As the Earth rotates, the line of radio telescopes traces out part of a ring in space, and the signals received by the various telescopes can be combined to produce the resolving power of an instrument with a diameter equal to the separation of the widest pair of dishes. Since this process produces a resolution equivalent to one very large telescope, the technique is often referred to as 'aperture synthesis'.

A refinement of the basic aperture synthesis technique is to have some of the telescopes mounted on rails so that the distance between the dishes can be varied from day to day, making it possible to produce very detailed maps of radio sources. An example of such a telescope is the 5 km (3 mile) telescope at Cambridge, England, which, depending on the wavelength under study, can produce radio maps with a resolution of 0.01 arc minutes. This performance makes the Cambridge radio maps comparable with photographs taken using a good optical telescope. The largest

---

[†]The resolving power of a telescope is given by $1.22\lambda/D$, where λ is the wavelength of the radiation and D is the diameter of the telescope. The long wavelengths of interest to radio astronomers means that a radio telescope with the resolving power of an optical telescope would be impossibly large.

aperture synthesis telescope is the Very Large Array (VLA) in Sirroco, New Mexico, which combines 27 radio dishes, each 26 m in diameter, in a Y shaped pattern with arms 21 km long, into a single instrument.

The radio telescopes used for aperture synthesis are usually located reasonably close together so that the signals from each telescope can be sent directly, either by waveguides or microwave links, to a central control room for recording. There is, however, no reason, other than the difficulties of communications and data processing, why telescopes separated by much greater distances cannot be combined to provide a resolving power equivalent to that of a very large instrument. In this case it is usual to record the signals received at each telescope, together with a reference signal based on accurate atomic clocks, at each observatory rather than transmitting them directly to a central station. At the end of each series of observations the tape recordings are sent to a single laboratory where, using the reference signals provided by the atomic clocks, the recordings are combined and the results extracted. The technique, known as Very Long Baseline Interferometry (VLBI), can be applied to telescopes separated by enormous distances, but is obviously limited to the diameter of the Earth. A further limitation of ground based VLBI is that since large radio telescopes cannot be moved easily across great distances, both the size and orientations of the VLBI baselines available to study a given astronomical object are fixed by the geometry of the Earth. This problem is compounded by the distribution of radio telescopes; there are far more in the Northern hemisphere than in the Southern hemisphere.

An orbiting radio telescope dedicated to VLBI observations in conjunction with ground based radio telescopes offers the promise of improvements in several areas. Firstly, longer baselines and hence even greater angular resolution than any two terrestrial observatories; secondly, the constant motion of the satellite as it travels around the Earth means that the lengths and orientations of the available baselines can be varied; and thirdly, a satellite in a highly inclined orbit can provide improved coverage of objects in the southern sky. These benefits mean that radio astronomers are now actively promoting the idea of space VLBI missions.

7.3.1 Experiments with the TDRSS satellite

Experiments to prove the feasibility of space VLBI were attempted with the Salyut 6 KRT-10 instrument (see section 7.2.4) but better known are those made with one of the American Tracking and Data Relay Satellite System (TDRSS) communications satellites. The TDRSS system, which is based on the concept of several large spacecraft in geostationary orbit, is designed to provide a near continuous communications link with low flying satellites and so reduce the number of ground stations which NASA needs to maintain. The fully operational system will consist of two satellites (plus one spare), but by early 1988 only one of these had been launched, the second being destroyed in the explosion of the Space Shuttle Challenger in January 1986. As part of their communications payload the TDRSS satellites are equipped with two large (4.9 m diameter) antennae, and this makes them suitable, under certain circumstances, for space VLBI experiments. The first such observations were made in July and August 1986.

The TDRSS satellites are designed to look down on other spacecraft and so are restricted to pointing in the general direction of the Earth. This operational limit, which would not apply to a dedicated VLBI satellite, severely restricted both the objects which could be observed during the experiment and the choice of radio telescopes which could serve as the other elements in the VLBI array. The two radio dishes best suited for the experiment were a 64 m diameter dish at Tidbinbilla in Australia, which forms part of the NASA deep space tracking network, and a similar sized radio telescope at Usuda in Japan. A smaller, 26 m, radio telescope in Japan was used during most of the observations to verify the performance of the larger telescopes.

Both of the 4.9 m antennae on the TDRSS satellites were employed during the VLBI observations, one to record the radio signals from the object under study, the other to monitor a beacon signal transmitted from the TDRSS ground station at White Sands, New Mexico. The purpose of the beacon signal was to provide information on any motion of the satellite relative to the other radio telescopes during the observations. Data from the satellite were relayed to the ground and recorded along with a suitable reference signal. These data were later correlated with tape recorded data from the other radio telescopes in the usual way.

During the first phase of the experiments, several quasars were observed at a wavelength of 13 cm (2.3 GHz). The longest baseline achieved was 1.4 Earth diameters, this being the effective distance between the TDRSS spacecraft and the radio telescope at Usuda during observations of the quasar known as 1730–130. Data from this observation, and several others which produced rather smaller baselines, were sucessfully combined to produce useful astronomical measurements, demonstrating the feasibility of space based VLBI missions.

7.3.2 Future Space VLBI Missions

Several space VLBI missions have been proposed, but it is not yet clear which, if any, of these will finally take place. Since VLBI is, by its very nature, an international undertaking, it is almost certain that the various proposed missions will be merged to some extent, either as a single multinational project or as coordinated programme of separate satellites. In view of the doubts about which satellites will eventually be launched, only brief details of each proposal will be given here.

RACSAS (Radio Astronomic Cosmic System of Aperture Synthesis) is a Soviet proposal for a satellite to be placed in low Earth orbit. The space radio telescope would be 30 m in diameter and be placed in an orbit 4–500 km high inclined at 50° to the Equator. In such an orbit the satellite would complete several revolutions per day and so provide many different baselines between itself and ground based radio telescopes every few days. This would allow detailed maps of compact radio sources to be built up from scans at different orientations, but would not provide greatly improved angular resolution because the maximum baseline possible would be only just over one Earth diameter. Data from the low flying satellite would be relayed to a ground station via a relay satellite in a higher orbit.

Radioastron is also a Soviet proposal, but unlike RACSAS it is aimed at providing baselines much longer than those possible from low Earth orbit. In the initial phase of the Radioastron project a satellite equipped with a 10 m diameter dish would be placed in a highly elliptical orbit reaching an apogee close to 100 000 km. The inclination of the orbit would probably be about 60°. Later versions might be placed into orbits with an apogee of 1 million kilometres to provide even longer baselines and hence still greater

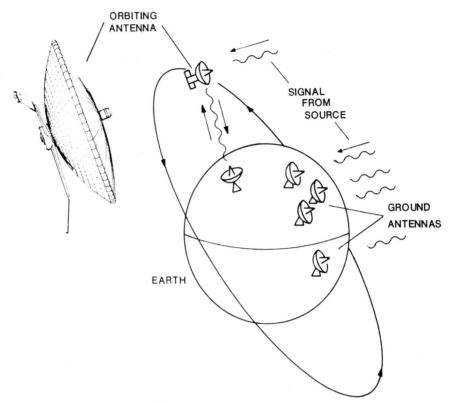

The principle of Space VLBI. (ESA).

angular resolution. The initial version of Radioastron may be launched in the mid 1990s.

The QUASAT (QUAsar SATellite) project was proposed in 1982 as a joint NASA–ESA space VLBI mission, and design studies had progressed quite far when the Challenger accident forced NASA to reconsider its space astronomy programme. As a result ESA re-evaluated QUASAT as an essentially European project with substantial involvement from the United States, Australia, and Canada, all countries with considerable expertise in ground based radio astronomy. QUASAT is conceived as a 15 m diameter radio telescope in an inclined, elliptical orbit which would collect data and transmit them to the ground immediately. The data would be recorded as they were received at the ground station and eventually passed to a central laboratory to be merged with data from cooperating ground based telescopes.

Various orbits have been considered for QUASAT. One of these stretches from 5 700–12 500 km and is inclined at 63° to the Equator, allowing high quality mapping in conjunction with US and European ground based telescopes. An alternative, which offers lower launch costs and improved angular resolution although is less suited for mapping, is from 5 000–36 000 km and inclined at 30°. The possibility of operating in the higher orbit for a time, then dropping into a lower orbit to provide better conditions for mapping, has also been considered. If selected for development, QUASAT would be launched about 1996.

Japanese proposals for space VLBI are at present based around a 10 m deployable antenna launched by a M-3SIII solid propellant rocket. This launch vehicle would allow a spacecraft weighing about half a tonne to be placed in an orbit reaching about 50 000 km, but the possibility of using a more powerful launch vehicle, able to reach a higher orbit, is being considered. The choice of launch vehicle is complicated by Japanese government policy which limits the maximum size of launch vehicle developed by ISAS, the Japanese organisation responsible for space science. The National Space Development Agency of Japan (NASDA) have more powerful rockets already available, but to date these have been used only for projects involving the development of new technology, never for scientific missions. The Japanese space VLBI satellite is also being regarded as a possible collaborative project with the European QUASAT mission.

8

The future

8.1 INTRODUCTION

Satellites already under construction, or expected to be developed when funds become available, have been described in Chapters 2 to 7. These projects provide a preview of space astronomy between the late 1980s and the closing years of the twentieth century. Beyond 2001 more powerful launchers and new instrumentation will combine to increase the power of orbiting telescopes far beyond that available today. The precise nature of the satellites of the next century is difficult to determine at present, and to give a catalogue of possible future missions, many of which will remain little more than ideas sketched on the back of the ubiquitous envelope, would be both tedious and pointless. Instead, this chapter will consider how the next generation of space telescopes may differ from those already described. In doing so it leaves the areas of description and prediction and enters the realm of speculation.

Many of the factors which will affect the next generation of astronomical satellites will not be connected with developments related to astronomy itself, but rather to the increasing exploitation of space for a variety of political, commercial, and military purposes. Dominant amongst these will be the increasingly easy access to space made possible by new generations of re-usable spacecraft and by the establishment of large, permanently manned space stations. Although the vast expenditures required to keep humans in space have been criticised by some scientists, many of whom fear that massive cost overruns in the space station project will result in the cutting back of smaller, but more productive unmanned scientific missions, there can be no doubt that the space station and its associated technologies offers new opportunities for astronomy. Areas of advanced technology which may be important to astronomy are the space station itself, the increased amount of traffic to and from Earth orbit, the in-orbit assembly of large structures, and the development of manned bases on the Moon.

8.2 THE SPACE STATION

The way in which a satellite mission is designed, with various choices and compromises being made to achieve a certain set of goals, was described in Chapter 1. The builders of astronomical instruments destined for space stations will not have many of these options; factors such as the orbit, communications rates, attitude control systems, and stability will be set by the station itself, not by the needs of individual payloads. In fact several aspects of the proposed international space station announced by President Reagan are quite unsuitable for astronomical missions. The space station orbit will be low, causing regular occultations of many astronomical sources by the Earth, the stability will be relatively poor, so that many astronomical experiments will need additional systems to achieve the required pointing accuracy, and the local environment will be too badly polluted, notably by water vapour and waste gases, for most cryogenically cooled instruments.

This is not to say that astronomy cannot be done from a space station; it may be possible to accommodate instruments which do not require very high levels of stability, which are relatively invulnerable to contamination, or which require large quantities of electric power. For example coded mask telescopes, for either gamma ray or X-ray astronomy, do not require a very stable mounting platform since the precise pointing direction of the instrument can be reconstructed during data processing. The structure of the space station will also allow the masks used in these instruments to be located some distance from the detector array, providing high angular resolution. Orbiting cosmic ray instruments, which to date have been severely limited in size because of their relatively high masses, are also suitable for mounting on the space station, especially if it is possible to devise modular units which can be brought up and assembled over a period of years into arrays too massive to be launched as a single instrument.

None the less, most types of astronomical instrument in use during the space station era will probably be installed on free-flying platforms able to operate away from the disturbing influences of human activity. Depending on their missions, the platforms may operate in low orbits near to the space station or may be boosted to greater altitudes for routine operations, returning to the space station only when maintenance is necessary. A major role of the space station in 21st century astronomy may thus be the provision of an orbiting facility where satellites can be repaired, resupplied with consumables (for example liquid helium for infrared telescopes, spark chamber gas for gamma ray instruments), or have new instruments fitted.

The free-flying platforms used to carry astronomical instruments will probably come in two forms: large platforms for long duration missions, and small, simple spacecraft for missions lasting for only a few days or weeks. Typical of a long lived platform is the European Retrievable Carrier (Eureca). The basic Eureca spacecraft is a 4 tonne platform, $2.3 \times 2.3 \times 4.5$ m in size, which is carried into orbit by the Space Shuttle. After release the Eureca boosts itself up to a higher orbit, probably about 500 km in altitude, and operates independently for several months. At the end of its mission the Eureca returns to low orbit and is recovered by the Space Shuttle and returned to Earth for refurbishment and relaunch. The first Eureca will be used for microgravity research, and so has only limited stability, but a follow on version could have improved pointing accuracy and could be used for astronomical missions. It is a fairly simple step to extend the Eureca concept to platforms which are equipped at a space station, carry out their mission, then return to the station to be fitted with new instruments.

The Spartan-1 satellite after deployment from the Space Shuttle. (NASA/NRL).

Smaller free flyers, fitted with payloads similiar to those carried by sounding rockets, are also possible and two such systems already exist. The NASA Spartan programme is based on simple, low cost platforms deployed from the Space Shuttle which carry out a mission lasting two or three days before being recovered and returned to Earth. Spartan satellites are able to operate clear of the environment around the Shuttle, and can point in any direction without the constraints imposed on payloads which remain attached to the Shuttle. The Spartan is little more than a framework, equipped with batteries, a gas attitude control system, and a powerful tape recorder, into which a single astronomical payload is fitted. Spartan-1 was deployed on 17 June 1985 from Space Shuttle mission 51G and was retrieved 45.5 hours later. During its two day flight the spacecraft's collimated X-ray detectors observed the Perseus Cluster of galaxies and the centre of our own Galaxy in the energy range 1–10 KeV. The next planned Spartan missions are expected to be devoted to solar physics and to stellar ultraviolet astronomy. The German Shuttle Pallet Satellite (SPAS) is similar in concept to the Spartan vehicle. The first SPAS was carried on the seventh Space Shuttle mission and was used to practise the techniques of satellite deployment and retreival. SPAS-01 did not carry astronomical instruments, but a second flight, carrying an extreme ultraviolet telescope, is planned and is known as ASTRO-SPAS.

Such vehicles have many of the advantages (for example low cost, simplicity) of sounding rockets, and yet offer much longer observing times. In the era of Space

The Spartan-1 satellite attached to the remote manipulator arm of the Space Shuttle. (NRL).

Stations and regular traffic to and from Earth orbit they offer an important opportunity for simple astronomical experiments. Similar short duration missions may be launched from and serviced at the Space Station itself to avoid the need to carry the basic spacecraft to and from the Earth every time.

8.3 IN-ORBIT ASSEMBLY

One of the advantages of space stations is the ability to use them as a base for the assembly of very large structures. In-orbit assembly offers two advantages: the final structure can be much larger than could be lifted by a single rocket, and there is no need to design the structure so that it can survive the pounding it receives during the first few minutes of launch. Structures designed to be assembled in space can be transported to orbit in suitable protective packaging and then constructed and operated in the much

184   The future

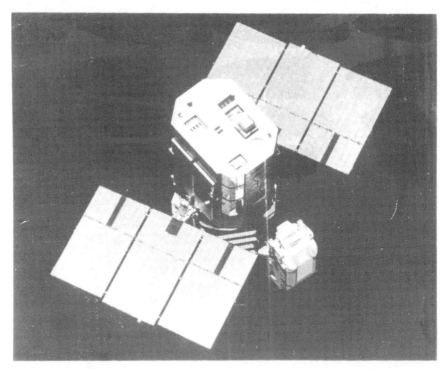

A space suited astronaut attempts to capture the ailing Solar Maximum Mission in 1984. In orbit assembly, maintenance and repairs will be routine for future large astronomy missions. (NASA).

more benign zero gravity environment of space. Several large astronomical instruments have been considered for in-orbit assembly, and two will be described here as examples of how the technology of the next century might be applied to astronomy.

GRITS is a Gamma Ray Imaging Telescope System designed to be assembled inside a Space Shuttle external fuel tank. The objective of the instrument is to provide a large gamma ray telescope with high sensitivity and good angular resolution which can follow up the results from the Gamma Ray Observatory sky survey. The instrument works in the following way: gamma rays enter one end of the tank and pass into a module where they are converted into electron–positron pairs as they encounter a sheet of metal. Next, a trigger module consisting of a sheet of plastic scintillator records the passage of the electron–positron pairs into the main volume, a vessel containing large quantity of gas. As the high speed electrons and positrons travel through the gas they exceed the local velocity of light within the gas and so emit flashes of light by the Cherenkov process. The visible photons are then collected by an optical system and recorded. The electron–positron pair eventually reaches the far end of the gas volume and encounter a final scintillation counter which records them as they escape from the far end of the telescope.

The GRITS telescope would have a collecting area of about 25 square metres, a field of view of a few degrees, and an angular resolution of a few arc minutes. Furthermore, since the time intervals between the trigger pulse, the departure of the electron–positron pair through the rear of the telescope, and the time of the flash of Cherenkov

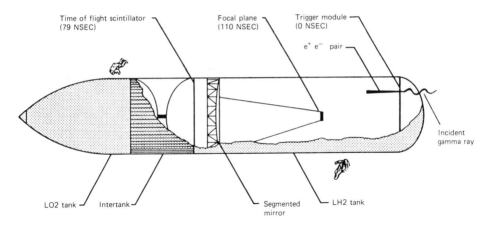

Diagram of the GRITS telescope. Times in brackets are times of detections following the registration of a gamma ray by the trigger detectors. (Martin Marietta).

radiation are fixed by the geometry of the instrument, GRITS can reject background events caused by charged particles, making it more sensitive to faint gamma ray sources.

The basic design of the GRITS telescope was proposed in 1966, and several balloon flights of a prototype were made in the early 1970s. The ideas then lay dormant for some years because there was no way of launching the large, rigid, gas and light tight structure required for a larger version of the instrument. However, the advent of in-orbit construction makes the idea viable because the Space Shuttle external tank, normally allowed to fall to destruction but quite capable of being carried into orbit if required, is an almost ideal structure into which GRITS could be fitted. All that would be required is for the empty tank to be brought to the space station where astronauts could clean it out, install the necessary optical and electrical systems together with an attitude control system, and pressurise the tank with a suitable gas. Such an operation would be virtually impossible today, but well within reach of space station astronauts of the next century.

A more ambitious proposal for in-orbit assembly involves coupling together four large mirrors, each 4 m in diameter, to form a 36 m long array of optical telescopes known as COSMIC (Coherent Optical System of Modular Imaging Collectors). The four mirrors comprise an optical interferometer with the resolving power of a much larger mirror. Although a prototype using four 1 m mirrors arranged in a 10 m array could be launched by a single Space Shuttle, the larger version would need to be assembled in orbit, using components brought up on several flights. The assembly of the mirror array, together with the addition of solar panels and sunshades to the structure, would be completed by astronauts based at the space station, removing the need for complex automatic assembly sequences. Since the optical paths from each of the mirrors to the centre of the array must be equal, and must be maintained with a precision of a fraction of the wavelength of light, further visits by astronauts to make adjustments to the alignment would probably be needed during the initial commissioning of the instrument. Ultimately, the size of the COSMIC telescope could be increased by adding further elements to produce a large, two-dimensional optical telescope array of prodigious power.

COSMIC is not the only multiple mirror system to be investigated. Other studies have considered the optical performance, mass, and assembly problems of phased optical arrays of various sizes, including an 18 element array of 3.2 m diameter mirrors which would have a total collecting area of 145 square metres and an overall diameter of 39 m. The assembly of such a structure would be a major task; the estimated time to assemble a smaller, 6 mirror, array considered during the same study amounted to almost 800 hours. These studies also considered the desirability of placing such a structure in geostationary orbit to take advantage of the improved sky coverage, fewer eclipses by the Earth, smaller gravity gradient torques on the structure, and reduced risk of damage by collisions with space debris, even though such a high orbit would incur increased problems of transportation and assembly. Much more work will be required before the design of such arrays can be optimised.

8.4 SATELLITE CLUSTERS

COSMIC and its close relatives are designed to be used as optical interferometers able to provide angular resolution greater than that achievable from a single telescope, but are limited by the need to join several mirrors together in a single structure. An alternative, which enables the elements of the interferometer to be located further apart and thus produce greater resolution, is to use several satellites flying in formation. Since an interferometer operating at optical wavelengths must maintain its various elements in position to an accuracy of a few hundredths of a micrometre, the use of a satellite cluster rather than a single rigid structure presents many technological problems, but these may not be insuperable as space technology continues to improve. If these problems can be overcome, then such a cluster offers an obvious advantage over a single structure; the elements can be separated by much greater, and infinitely variable, intervals.

Two studies of satellite clusters for interferometry have been made: the Spacecraft Array for Michelson Spatial Interferometry (SAMSI), conceived in the USA, and the European TRIO. SAMSI envisages two, 1 m, telescopes in orbits which take them up to 10 km apart. This separation corresponds to a resolving power of 10^{-5} arc second, 100 000 times better than a ground based optical telescope. The two telescopes, and a third spacecraft which serves to receive the beams from the telescopes and combine them, would be placed in equatorial orbits about 1 000 km high, but with slightly different inclinations. Major technical considerations are how to predict the correct position for the central receiving spacecraft and how to ensure that the receiver is maintained at that position. Since the orbits of all three satellites intersect, avoiding a collision is also an important design consideration. In the SAMSI proposal the satellites would be equipped either with low thrust ion engines or conventional chemical thrusters which could be refuelled at a space station to provide manoeuvring capability. The European TRIO proposal is broadly similar to SAMSI but uses the pressure of sunlight on solar sails as a means of rotating the satellites and moving them relative to each other. Such a system is virtually vibration free and is not dependent on limited supplies of fuel. The TRIO cluster might be located at one of the Lagrangian points, regions of space where differential gravitational forces are weak and small satellites can remain in relative equilibrium for long periods.

Another application of satellite clusters is for very high angular resolution coded mask telescopes. Since the resolution of coded mask telescope varies with the separation of the mask and the detector there is a possibility of mounting a gamma or X-ray detector on one satellite and a suitable coded mask on another co-orbiting platform. With accurate pointing of the two satellites, very high resolution gamma and X-ray imaging is theoretically possible.

8.5 ASTRONOMY FROM THE MOON

Although from the vantage point of the late 1980s the prospect of a permanent lunar base appears remote, there seems little doubt that such a base will one day be constructed. The reasons for such an undertaking will be many and complex and will not be dominated by the desires of astronomers, but when such a base is available it will clearly have potential for astronomy. A few instruments which might be built on the Moon will be mentioned here. Detailed technical studies can be found amongst the scientific journals.

The most obvious advantage of the lunar surface for astronomy is the lack of atmosphere. This would allow observations over the full range of the electromagnetic spectrum and diffraction limited imaging, unaffected by atmospheric turbulence, at optical wavelengths. These opportunities also exist for Earth satellites, but lunar based facilities offer considerable advantages in terms of pointing stability and ease of maintenance. A lunar observatory also offers long uninterrupted pointings of up to 14 Earth days during the lunar night and the potential for constructing passively cooled infrared telescopes able to operate indefinitely at very low temperatures. The low lunar gravity also makes large telescopes, both optical and radio, easier to construct and less vulnerable to gravitational distortion than their terrestrial counterparts.

Instruments placed on the Moon during the Apollo programme have shown that the lunar surface is very stable, so structures built on the Moon will remain fixed in their relative positions and orientations for long periods. This, together with the absence of scintillation due to an atmosphere, makes the Moon an ideal site for long baseline optical interferometry. It is possible to imagine that the next century will see the establishment of a lunar Very Large Array consisting, not of radio telescopes as in the terrestrial case, but many optical telescopes linked together. Detailed designs of such a structure have not yet been made, but one study published in 1984 suggested an array of 27 telescopes, each 1 metre diameter, arranged in a Y with arms about 6 km long.

The Moon also offers exciting prospects for radio astronomy. One of these is the possibility of extending existing Very Long Baseline Interferometry techniques to a baseline stretching from the Earth to the Moon. A Moon Earth Radio Interferometer (MERI) operating at a wavelength of 6 cm could produce radio images with a resolution of 30 micro-arc seconds, many times greater than terrestrial VLBI networks. Since a lunar base will need at least one large dish antenna for routine communications with the Earth, it may be possible to begin MERI experiments with this dish and then to construct a dedicated radio astronomy dish once the basic ideas had been proved. A lunar VLBI antenna would have a much longer lifetime than an orbiting radio telescope of the QUASAT/Radioastron type, although the distance between the lunar and terrestrial dishes could not be varied in the same way as is possible with an orbiting dish. For improved coverage MERI observations might be combined with data from a QUASAT-like orbiting telescope.

Lunar radio telescopes might also be used to extend the work of the Radio Astronomy Explorer spacecraft by observing at very low frequencies (VLF) impossible from the Earth. The Moon is an ideal platform for such an observatory since large numbers of antenna can be laid out in positions which will not move relative to each other, and the properties of the lunar soil mean that antennae can be laid out along the ground; there is no need to construct or maintain structures to support them. Other advantages are that the array can be extended at any time, even during journeys undertaken for other purposes, and that lunar rotation would provide a monthly scan of the sky. An array many kilometres on a side could produce an angular resolution as high as one degree, twenty times better than the precision of the RAE spacecraft. The lunar VLF observatory would be best placed on the far side of the Moon where it would be shielded from the radio emissions from the Earth, but logistical considerations would probably mean that it would be laid out close to wherever the first lunar base was established.

8.6 CONCLUSION

This very brief chapter has shown how astronomers might take advantage of improved space technology to develop new instruments able to study the Universe in hitherto unachievable detail. Some of these instruments may be built during the early part of the next century, and others, using ideas that have yet to be put on paper, certainly will. What these new telescopes will discover cannot be foreseen, but it is hoped that some of the readers of this book, and perhaps even its author, will one day use them to improve mankind's understanding of the Universe. Although satellites have done much to uncover the secrets of the sky, there is much that remains hidden.

Bibliography

This bibliography is not intended as a complete list of sources used in the preparation of this book, nor is it a comprehensive review of the available literature. Instead, it lists a selection of material which includes and expands on what is written here and which may be of some value to the reader. The first section assumes only an interest in, and general understanding of, astronomy. The selection of technical works is aimed at the more experienced amateur and at professional scientists. Further information can be found by using the references within the literature cited here, and by the use of the usual citation indexes.

FOR THE GENERAL READER

BOOKS

Allen, D. A. *Infrared, the new astronomy*. Keith Reid Ltd, Shaldon, Devon, 1985.
Bester, A. *The life and death of a satellite*. Sidgwick & Jackson, London, 1967.
Clark, D. *The quest for SS433*. Adam Hilger, Bristol, 1986.
Cooper, S. F. *A house in space*. Angus & Robertson (UK), London 1977.
Gatland, K. et al. *The illustrated encyclopedia of space technology*. Salamander Books, London, 1981.
Greenstein, G. *Frozen star*. Macdonald & Co, London, 1983.
Henbest, N. & Marten, M. *The new astronomy*. Cambridge University Press, 1983.
Hirsh, R. *Glimpsing an invisible universe*. Cambridge University Press, 1983.
Kopal, Z. *Telescopes in space*. Faber & Faber, London, 1968.
Lundquist, C. A. (ed.). *Skylab's astronomy and space sciences*. NASA SP-404, NASA Washington DC 1975.
Moore, P. et al. *The astronomy encyclopedia*. Mitchell Beazley, London, 1986.
Sharpe, M. *Satellites and probes*. Aldus Books, London, 1970.
Tucker, W. *The star splitters*. NASA SP-466. Washington DC, 1984.
Tucker, W. & Tucker, K. *The cosmic enquirers*. Harvard University Press, Cambridge Mass 1986.

Tucker, W. & Giacconi, R. *The X-ray Universe*. Harvard University Press, Cambridge Mass, 1985.
NB. There is considerable overlap between the three books by W. Tucker.

Magazine Articles

'Astronomy through the Skylab scientific airlocks.' *Sky and Telescope* May 1973.
'Notes on Soviet space astronomy.' *Sky and Telescope* February 1977.
'Astronomy with Salyut 6.' *Sky and Telescope* January 1982.
'Astronomy from Spacelab-1.' *Sky and Telescope* July 1984.
'Pioneering balloon astronomy in France.' *Sky and Telescope* 1983.
'The dawn of balloon astronomy.' *Sky and Telescope* December 1986.
'KAO, mission incredible.' *Astronomy* June 1980.
'NASA's 91 cm airborne telescope.' *Sky and Telescope* November 1986.
'The March eclipse rocket program at Wallops Island.' *Sky and Telescope* June 1970.
'Sounding rockets in space astronomy.' *Sky and Telescope* October 1972.
'The first X-ray astronomy satellite.' *Sky and Telescope* January 1971.
'The OSO-7 year of discovery.' *Sky and Telescope* January 1973.
'X-ray Astronomy with HEAO-1.' *Sky and Telescope* December 1978.
'The X-ray eyes of Einstein.' *Sky and Telescope* June 1979.
'EXOSAT: Europe's new X-ray satellite.' *Sky and Telescope* May 1984.
'Stellar coronas, X-rays and Einstein,' *Sky and Telescope* July 1984.
'Spacelab-2: science in orbit.' *Sky and Telescope* November 1986.
'The high flying Kvant module.' *Sky and Telescope* December 1987.
'Gamma rays—the last frontier.' *Mercury* May–June 1981.
'Gamma ray astronomy comes of age.' *Sky and Telescope* October 1985.
'Another orbiting astronomical observatory.' *Sky and Telescope* December 1970.
'An astronomy satellite named Copernicus.' *Sky and Telescope* October 1972.
'Ultraviolet astronomy with the satellite Copernicus.' *Mercury* 1976.
'The IUE satellite.' *Sky and Telescope* December 1973.
'Observing with IUE.' *Mercury* March–April 1979.
'The best little telescope in the solar system' (IUE). *Space World* January 1987.
'An ultraviolet eye on the sky.' (IUE) *New Scientist* 27 January 1987.
'IUE, nine years of astronomy.' *Astronomy* April 1987.
'The extreme ultraviolet.' *Astronomy* July 1987.
'Space telescope, eye on the Universe.' *Spaceflight* December 1982.
Space telescope special issue. *Sky and Telescope* April 1985.
'HST, astronomy's greatest gambit.' *Sky and Telescope* May 1985.
'Building the space telescope's optical system.' *Astronomy* January 1986.
'Catch a hundred thousand stars.' (Hipparcos). *New Scientist* 1 July 1982.
'The frigid world of IRAS'. *Sky and Telescope* January and February 1984.
'IRAS and the infrared Universe.' *Astronomy* March 1984.
'Infrared eyes on the Universe.' *Space World* February 1987.
'COBE'S quest.' *Astronomy* August 1986.
'Astronomy from satellite clusters.' *Sky and Telescope* March 1984.

The above magazines are published by the following organisations and provide a means of keeping in touch with the rapidly developing world of space astronomy.

Astronomy: Astromedia Corp/Kalmbach Publishing, 1027 N Seventh St, Milwaukee, USA.
Mercury: Astronomical Society of the Pacific, 1290 24th Av, San Fransisco, CA 94122, USA.
New Scientist: New Science Publications, Commonwealth House, 1–19 New Oxford St, London WC1A 1NG.
Sky and Telescope: Sky Publishing, 49 Bay St, Cambridge, Mass, USA.
Spaceflight: British Interplanetary Society, 27/29 South Lambeth Rd, London.
Space World: National Space Society, 922 Pennsylvania Av SE, Washinton DC, USA.

TECHNICAL READING

Books

Culhane, J. L. and Sanford, P. W. *X-ray astronomy*. Faber and Faber, London, 1981.
Kondo, Y. (ed.) *Exploring the Universe with IUE*. D Reidel, Dordrecht, Holland. 1987.
Ramana Murthy, P. V. and Wolfendale, A. W. *Gamma-ray astronomy*. Cambridge University Press, 1986.
Hillier R. *Gamma ray astronomy*. Clarendon Press, Oxford, 1984.

Journals

Balloon and rocket astronomy

Friedman, H. 'Rocket astronomy.' *Scientific American* **200 (6)**: 52–60 1959.
McCarthy, D. J. 'Operating characteristics of the Stratoscope II balloon-borne telescope.' *IEEE Transactions on Aerospace and Electronic Systems*. **AES-5 (2)** 323–329 1969.

Gamma ray astronomy

Agrinier, B. *et al.*, 'Gamma 1: a telescope for 50–5000 MeV astronomy.' *Soviet Astronomy* **30(5)**: 508–514 (in English).
Bennett, K. 'COS-B, A mission fully accomplished.' *ESA Bulletin* **45**: 40–43 1986.
Dean, A. J. 'Future space missions in gamma-ray astronomy.' *J. British Interplanetary Soc.*: 1988.
Fitchel, C. E. 'Gamma ray astrophysics.' In: Bernacca, P. and Ruffini, R. (eds). *Astrophysics from Spacelab*, D Reidel Publishing 1980.
Fitchel, C. E. *et al.*, 'High energy gamma ray results from the second small astronomy satellite.' *Ap. J.* **198**: 163–182 1975.
Lust, R. 'Gamma ray astronomy and the spirit of COS-B.' *ESA Bulletin* **42**: 8–16 1985.
Mandrou, P. 'The Sigma mission.' *Advances in Space Research* **3** :525–531, 1984.
Taylor, B. G. and Willis R. D. 'Six years of gamma ray astronomy with COS-B.' *ESA Bulletin* **28**: 48–61 1981.
Wolfendale. A. W. 'Gamma ray astronomy.' *Q. J. R. Astr. Soc.* **24**: 226–245 1983.

X-Ray and EUV astronomy

Anon. 'The high energy astronomical observatory programme.' *J. British Interplanetary Soc.*: **31** 424–431 1978.
Beigman, I. L. *et al.* 'Observation of an X-ray flare near lambda Scorpii.' *Soviet Astronomy Letters* **2(1)**: 6–8 1976 (in English, describes the Salyut-4 X-ray telescope).
Bratolyubova–Tsulukidze, L. S. *et al.* 'Kosmos 428 observations of hard X-rays from the galactic-center region.' *Soviet Astronomy Letters* **2(1)**: 4–6 1976 (in English).
Courtier, G. M. 'UK involvement in the ROSAT project.' *J. British Interplanetary Soc.*: **39** 217–223 1986.
Davies J. K. (ed.) 'Space science at Birmingham University.' *J. British Interplanetary Society*: **40(4)**. 1986. This issue describes the Spacelab-2 XRT, coded mask imaging theory, ROSAT, and the Solar Maximum Mission.
Finocchiaro G. *et al.* 'Satellite for X-ray astronomy' (SAX). Presented at the *38th Congress of the IAF*. Brighton, England 1987.
Giacconi, R. 'The Einstein Observatory. New perspectives in astronomy.' *Science* **209**: 865–876, 1980.
Hoff, H. 'EXOSAT—the new extrasolar X-ray observatory.' *J. British Interplanetary Soc.*: **36** 363–367, 1983.
NASA. 'Apollo–Soyuz test project.' *Summary Science Report* Volume **1**. NASA SP-412.

Taylor, B. G. et al. 'The European X-ray observatory satellite-EXOSAT.' *ESA Journal* **6**: 1–19 1982.
Terrell, J. 'The Vela 5B spacecraft: long-term X-ray astronomy.' *J. British Interplanetary Soc.*: 1988.
Trumper, J. 'ROSAT.' *Physica Scripta* **T7**: 209–215 1984.
Ulmer, M. et al. 'Spartan-1 X-ray observations of the Perseus cluster., *Ap J.*: **319** 118–125 1987.

Ultraviolet astronomy

Bogess, A. et al. 'The IUE spacecraft and instrumentation.' *Nature* **275**: 372–385 1978.
Boyarchuk, A. A. et al. 'The ultraviolet telescope aboard the astrophysical space station ASTRON.' *Soviet Astronomy Letters* **10(2)** 67–72 1984.
Cash W. 'The far ultraviolet spectroscopic explorer.' *J. British Interplanetary Soc.*: **37** 81–85 1984.
Carruthers, G. R. 'Apollo 16 far ultraviolet camera/spectrograph: instrument and operations. *Applied Optics* **12(10)**: 2501–2508 1973.
Severny, A. B. and Zvereva A. M. 'A possible interpretation of the UV background radiation observed with the space experiment GALAKTICA.' *Astrop Lett.* **23**: 71–75 1983.
van Duinen, R. J. 'The ultraviolet experiment onboard the astronomical Netherlands satellite–ANS.' *Astron. & Astrop.* **39**: 159–163 1975.

Optical astronomy

McRoberts, J. 'Space telescope.' *NASA EP-166* NASA Washington DC.
Macchetto, F. et al. *The Faint Object Camera for the Space Telescope.* **ESA SP-1028** October 1980.
Bachall, J. 'The space telescope.' *Scientific American* **247(4)**: 40–51 1980.
Bouffard, M. 'HIPPARCOS.' *23rd Goddard Memorial Symposium.* NASA Goddard Spaceflight Center 27–29 March 1985.

Infrared astronomy

Anon. 'IRAS, the infrared astronomical satellite.' *Nature* **303** (5915) 287–291 1983.
Beichman, C. 'The IRAS view of the Galaxy and the solar system.' *Ann. Rev. Astron. Astrop.* **25**: 251–264 1987.
Holdaway, R. (ed.). 'Infrared astronomical satellite. *J. British Interplanetary Soc.* **36(1)**: 1–48. A special issue devoted to the IRAS mission and its ground operations system.
Koch, D. et al. 'Infrared telescope on Spacelab-2.' *Optical Engineering* **21**: 141–147 1981.
Lemke, D. et al. 'GIRL, the German infrared laboratory for spacelab.' *Advances in Space Research* **2(4)**: 123–130 1983.
Mather, J. C. 'The cosmic background explorer.' *SPIE* **280**: 20–28 1981.
Price, S. D. et al. 'Airforce Geophysics Laboratory infrared sky survey experiments.' *SPIE* **280**: 33–43 1981.
Rieke, G. H. et al. 'Infrared astronomy after IRAS.' *Science* **231**: 807–814, 21 February 1986.
Sholomitski, G. B. et al. 'A cooled submillimeter telescope.' *Soviet Astronomy* **30**: (5) 514–518 1987 (In English–describes the Aelita mission.)
Soifer, B. et al. 'The IRAS view of the extragalactic sky.' *Ann. Rev. Astron. Astrop.* **25**: 187–230 1987.
Strukov, I. A. and Skulachev, D. P. 'Deep space measurements of the microwave background anisotropy: first results of the RELIKT experiment.' *Soviet Astronomy Letters* **10(1)**: 1–4 1984 (In English).

Radio astronomy

Alexander, J. K. 'New results and techniques in space radio astronomy.' In Labuhn & Lust (eds). *New Techniques in Space Astronomy* 401–418. 1971.

Alexander, J. K. 'Scientific instrumentation of the Radio-Astronomy-Explorer-2 satellite. *Astron & Astrop* **40**: 365–371 1975.

Andreyanov, V. V. *et al.* 'Project Radioastron: an Earth-space interferometer.' *Soviet Astronomy* **30(5)**: 504–508 1987 (in English).

Brown, L. 'The galactic radio spectrum between 130 and 2600 kHz.' *Ap. J.* **180**: 359–370 1973.

Sagdeev, R. 'Some prospects of space VLBI. *Proc. Workshop on QUASAT. Austria June 1984.* ESA SP-213 Sept 1984.

Weber, R. R. *et al.* 'The Radio Astronomy Explorer Satellite, a low-frequency observatory.' *Radio Science* **6**: 1085–1097, 1971.

Weiler, K. W. *Radio Astronomy from Space Proceedings of a workshop at NRAO, Green Bank W Virginia USA 1986.*

Future missions

Bonnet, R. M. 'Advanced space instrumentation in astronomy.' *Advances in Space Research* **2(4)**. 1982.

Davies, J. 'Astronomy from the space station.' *J. British Interplanetary Soc.* **39**: 51–56 1986.

Longdon, N. & David, V. *ESA Workshop on Optical Interferometry in Space.* **ESA SP-273**, ESTEC, Nordwick, the Netherlands. 1987.

Mendell, W. (ed.) *Lunar Bases and Space Activities of the 21st Century.* Planetary Science Institute, Tucson, Arizona USA. 1986.

Index

1964–83C, 96
3C 273, 86
active collimator, 74
active cooling, 26
advanced OSO, 54
Aelita, 166
Aerobee rocket, 8
AFGL, 145, 146
airborne astronomy, 6
angular size of Earth, 22, 23
ANS, 43, 109
anticoincidence shield, 32
aperture synthesis, 175
Apollo 16, 81, 105
Apollo Telescope Mount, 54
Apollo–Soyuz mission, 121
Ariel 5, 45
Ariel 6, 59
ART-P, 90
ART-S, 90
Aryabhata, 49
AS&E, 33
ASTP, 121
ASTRO-1, 119
ASTRO-B, 59
ASTRO-C, 68, 69
ASTRO-SPAS, 182
astrometry, 133, 137, 141
ASTRON 1, 117
atmospheric absorption, 3, 168
atmospheric drag, 23
AXAF, 70

background radiation, 19
balloon-borne astronomy, 4, 72
BATSE, 92
big bang, 162
Black Brant rocket, 8
bolometer, 145
Bragg crystal spectrometer, 46
bremsstrahlung, 30
BST-1M, 166
Burnight T.R., 8

Caravane collaboration, 85
Celescope, 98
Cen X-3, 39
Cen X-mas, 47
Centaurus A, 43, 50, 58
Cherenkov detector, 84, 184
chopped photometric channel, 149
circular variable filter, 160
closed loop cooling, 27
COBE, 163
coded mask telescope, 65, 89, 92, 181, 186
collimator, 39
comets, 115, 153, 154
COMIS, 67
communications, 17, 18, 19
COMPTEL, 92
Compton scattering, 74
contamination (of instruments etc.), 23, 25, 27, 152, 158, 181
Copernicus, *see* OAO-3
Coronascope I, 4

Index

Coronascope II, 5
CORSA-B, 59
COS-A, 85
COS-B, 85
Cosmic Background Explorer, 163
cosmic background radiation, 162
cosmic ray, 72, 76
cosmic rays, 181
COSMIC, 185
Cosmos 51, 96
Cosmos 208, 36, 77
Cosmos 215, 36, 96
Cosmos 262, 36, 96
Cosmos 264, 77
Cosmos 428, 36, 41, 77
Cosmos 561, 77
Cosmos 731, 77
Cosmos 856, 87
Cosmos 914, 87
Cosmos 1106, 87
CPC, 149
Cyg X-1, 40, 42, 47, 57
Cyg X-3, 42, 86
Cygnus superbubble, 52

D2B-AURA, 109, 110
D2B-GAMMA, 110
DAX, 149
diffuse gamma ray emission, 86
diffuse X-ray background, 33, 42, 43, 50, 51, 52, 58, 110, 123
DIRBE, 163
DISK, 89
DMR, 165
Dollfus A., 4
doping, 145
dutch additional experiment, 149

EGRET, 92
Einstein observatory (naming of), 57
Einstein observatory, see HEAO-2
electrical model, 13
electronographic camera, 105, 107, 120
EMC testing, 14
energy calorimeter, 76
ESRO-2B, 35
EURECA, 181
EUV Explorer, 123
EUV, 94, 120, 121
EXOSAT X-ray telescopes, 63
EXOSAT, 61
Explorer 11, 76
Explorer 34, 81
Explorer 38, 169
Explorer 43, 172
Explorer 49, 172

Explorer 53, 49
Extrasolar planets, 138, 154

facility class satellite, 12
Faint Object Camera, 133, 135
Faint Object Spectrograph, 132
far ultraviolet, 94
FAUST, 118
Filin, 48
FIRAS, 165
FIRST, 167
flare star, 44, 123
flight model, 16
FOC, 133, 135
FOS, 132
free flying platforms, 181
free-bound radiation, 30
free-free radiation, see bremsstralung
Friedman H., 32
FUSE, 120, 121
FUV, 94, 120

galactic radio background, 172
galactic radio emission, 175
galactika UV expt, 110
gamma ray burst detector, 69, 81
gamma ray bursts, 78
gamma ray observatory, 91, 184
gamma ray production, 72
gamma ray spectrometer, 74
gamma ray spectroscopy, 87
GAMMA-1, 82, 88
gas scintillation proportional counter, 60, 67
geiger counter, 32
Geminga, 86
geocorona, 19, 20, 105, 110
Giaconni R., 33, 54, 134
GINGA, 68, 69
GIRL, 158
GLAZAR, 118
GRANAT, 82, 89
GRASP, 92
gravity gradient stabilisation, 171, 186
grazing incidence reflection, 53
grazing incidence telescope, 53, 120, 121, 123, 124
GRITS, 184
GRO, 91, 184
GSPC, 60, 67
guaranteed time, 29
guest observers, 29

HAKUCHO, 59
HEAO programme, 11, 50, 54
HEAO-1, 51, 82
HEAO-2 telescope, 54, 55
HEAO-3, 87

Index

HELIOS-2, 82
HELOS, 61
Her X-1, 40, 48
HEXE, 67
high speed photometer, 132
high throughput X-ray spectroscopy mission, 70
HIMS, 136
Hipparcos, 137
HISTAR, 146
Hopkins Ultraviolet Telescope, 120
hours confirmation, 150, 157
HRS, 132
HSP, 132
HST, 27, 118, 128
HST instruments, 131
HST operations, 133
HST servicing, 135
human eye (response of), 1
HUT, 120
HZ43, 122

image dissector, 132, 140
IMP-6, 81
in-orbit assembly, 183
infrared cirrus, 155
Infrared Space Observatory, 158
International Ultraviolet Explorer, 110
ionosphere, 168
IRAS, 146
ISEE-3, 81
ISO, 158
ISOCAM, 159
ISOPHOT, 160
IUE, 110

Janseen J., 4
Japanese satellites, 59, 69, 179
Jupiter, 115, 122, 171

KONUS, 81, 90
KRT-10, 175, 176
Kuiper Airborne Observatory, 7
KVANT, 67, 118

Lacroute P., 138
Large Astronomical Satellite, 110
Large Deployable Reflector, 167
Large Space Telescope, 127
LDR, 167
lead sulphide crystal, 143, 145
line emission, 31
LMC, 58
Low F., 145, 154
Low Resolution Spectrometer, 149
LRS, 149
lunar observatory, 106, 187
LYMAN, 120, 121

Magellan, 120
magnetorquers, 24
MERI, 187
Michelson interferometer, 165
microchannel plate, 56
Mir, 67, 118

Nauka module, 78
Naval Research Laboratory, 8, 32, 52
Neugabauer G., 145, 147
NICMOS, 135
Nova Monoceros, 47
novae, 116
NRL, 8, 32, 52

OAO programme, 96
OAO-1, 35, 97
OAO-2, 97, 98
OAO-3, 42, 100, 121
OAO-A2, 98
OAO-B, 99
OAO-C, 100
observatory satellite, 12, 18
OGO-3, 81
OGO-5, 76, 78
open loop cooling, 26
optical interferometry, 186, 187
Orbiting Astronomical Observatory, see OAO
Orbiting Solar Observatory, see OSO
Orion-1, 109
Orion-2, 108
OSO programme, 13, 34
OSO-3, 35, 76
OSO-7, 42, 81
OSO-8, 45, 50
OSO-C, 35
OSSE, 92
ozone, 94, 95, 99

pair production, 75
parallax, 138
parsec, 138
passive cooling, 26
peer review, 28
phase A, 13
phase B, 13
phase C/D, 13
Phobos mission, 82
Phoebus, 91
photometer, 96, 98, 109
Piccard A., 4
Pioneer Venus Orbiter, 81
plasma, 30
PODSOLNUKH, 90
PROGNOZ 6, 82, 110
PROGNOZ 7, 82
PROGNOZ 9, 82, 162

Index

PROGNOZ satellite, 78
proportional counter, 32
protoflight model, 17, 51
Proxima Centuari, 122
pseudo-random mask, 65
PULSAR X-1, 68
PULSAR X-2, 89

QUASAT, 178, 187
quasi periodic oscillations, 64

RACSAS, 177
Radio Astronomy Explorer, see RAE
Radioastron, 177, 187
RAE-1, 169, 188
RAE-2, 172
rapid burster, 49
reaction wheels, 24
RELIKT-1, 162
RELIKT-2, 163
RENTGEN, 67
resolving power, 175
RMC, 46
rockoons, 33
ROSAT X-ray telescope, 70
ROSAT, 27, 70, 124
rotation modulation collimator, 46
RS CVN stars, 64
RT-4, 48

Salyut-1, 109
Salyut-4, 48
Salyut-6, 166, 175
SAMSI, 186
San Marco platform, 39, 47
SAS-1, 38
SAS-2, 83
SAS-3, 45, 49
SAS-D, 111
satellite clusters, 186
SAX, 70
science team, 29
scientific planning, 28
scintillation counter, 73
Sco X-1, 48
seconds confirmation, 148, 157
semiconductor, 144
Seyfert galaxy, 48, 116
shadowgram, 65
SHEAL, 70
SIGMA, 89
SIGNE programme, 82, 110
SIRENE-2, 67
SIRTF, 161
SKR-02m, 117
sky coverage, 21
Skylab S019 UV experiment, 107

Skylab S150 X-ray experiment, 43
Skylab S183 UV experiment, 107
Skylab S201 UV experiment, 107
Skylark rocket, 8, 9, 32
slow rotator, 43
Small Astronomical Satellite, see SAS
SMM, 82, 124
SOFIA, 7
Solar Maximum Mission, 82, 124
solar UV spectra, 8
solar X-rays, 8, 31, 32, 33
South Atlantic Anomaly, 19
Soyuz 13, 108
Space Telescope, see HST
Spacelab-1 UV experiments, 118
Spacelab-2 IRT, 156
Spacelab-2 X-ray telescope, 66, 67
spark chamber, 75
SPARTAN, 182
SPAS, 182
spin stabilisation, 24
SS Cygni, 122
SS433, 88
starburst galaxies, 156
STIS, 136
Stratoscope I, 4
Stratoscope II, 5
structural model, 14
STScI, 134
Sun vector, 21
Super Explorer, 51
supernova 1987A, 6, 10, 68, 69, 116, 118, 120
survey missions, 21

TD-1A, 42, 103, 121
TDRSS, 19, 91, 134, 176
TENMA, 60
terrestrial radio noise, 171, 173
thermal control, 25
thermal testing, 15
Tousey R., 95
trigger detector, 75, 184
TRIO, 186
TTM, 67
TYCHO catalogue, 140

UHURU, 39
UIT, 120
Ultraviolet Astronomical Satellite, 110
ultraviolet glow, 24
Ultraviolet Imaging Telescope, 120
Ulysses, 82
US space station, 181
uvicon, 98

V2 rocket, 7, 95
Van Allen belts, 19, 20

Index

Vega, 154
Vela 5A/5B, 35, 81
Vela 6A/6B, 35, 81
Vela Hotel, 78
Vela satellites, 35, 78, 80
Venera probes, 81
VLBI, 175, 176, 187
Voyager, 121
VUV, 94

WATCH, 91
weeks confirmation, 150
WF/PC, 132, 135
White Sands missile range, 7, 33, 146

Wide Field Camera, 123, 124
Wide Field/Planetary Camera, 132, 135
Wolter H., 53
WUPPE, 119

X-ray bursters, 45, 47
X-ray Explorer, 38
X-ray Multi Mirror telescope, 70
X-ray Timing Explorer, 70, 124
XTE, 70, 124
XUV, 94, 121

zodiacal bands, 154

QB 136 .D38 1988

Davies, John Keith.

Satellite astronomy

UNIVERSITY OF ROCHESTER LIBRARIES

3 9087 01174506 6

NON-CIRCULATING PHYSICS
UNTIL

MAY 21 1990